Lecture Notes in Physics

T0171882

Volume 959

The Lecture Notes in Physics

The series Lecture Notes in Physics (LNP), founded in 1969, reports new developments in physics research and teaching-quickly and informally, but with a high quality and the explicit aim to summarize and communicate current knowledge in an accessible way. Books published in this series are conceived as bridging material between advanced graduate textbooks and the forefront of research and to serve three purposes:

- to be a compact and modern up-to-date source of reference on a well-defined topic
- to serve as an accessible introduction to the field to postgraduate students and nonspecialist researchers from related areas
- to be a source of advanced teaching material for specialized seminars, courses and schools

Both monographs and multi-author volumes will be considered for publication. Edited volumes should, however, consist of a very limited number of contributions only. Proceedings will not be considered for LNP.

Volumes published in LNP are disseminated both in print and in electronic formats, the electronic archive being available at springerlink.com. The series content is indexed, abstracted and referenced by many abstracting and information services, bibliographic networks, subscription agencies, library networks, and consortia.

Proposals should be sent to a member of the Editorial Board, or directly to the managing editor at Springer:

Lisa Scalone
Springer Nature
Physics Editorial Department
Tiergartenstrasse 17
69121 Heidelberg, Germany
Lisa.Scalone@springernature.com

More information about this series at http://www.springer.com/series/5304

Martin Bauer • Tilman Plehn

Yet Another Introduction to Dark Matter

The Particle Physics Approach

 Springer

Martin Bauer
Institut für Theoretische Physik
Universität Heidelberg
Heidelberg, Germany

Tilman Plehn
Institut für Theoretische Physik
Universität Heidelberg
Heidelberg, Germany

ISSN 0075-8450 ISSN 1616-6361 (electronic)
Lecture Notes in Physics
ISBN 978-3-030-16233-7 ISBN 978-3-030-16234-4 (eBook)
https://doi.org/10.1007/978-3-030-16234-4

This Springer imprint is published by the registered company Springer Nature Switzerland AG.
The registered company address is: Gewerbestrasse 11, 6330 Cham, Switzerland

Preface

As expected, this set of lecture notes is based on a course on dark matter at Heidelberg University. The course is co-taught by a theorist and an experimentalist, and these notes cover the theory half. Because there exists a large number of textbooks and lecture notes on the general topic of dark matter, the obvious question is why we bothered collecting these notes. The first answer is: because this was the best way for us to learn the topic. Collecting and reproducing a basic set of interesting calculations and arguments is the way to learn physics. The only difference between student and faculty is that the latter get to perform their learning curve in front of an audience. The second answer is that we wanted to specifically organize material on weakly interacting dark matter candidates with the focus on four key measurements:

1. Current relic density
2. Indirect searches
3. Direct searches
4. LHC searches

All of those aspects can be understood using the language of theoretical particle physics. This implies that we will mostly talk about particle relics in terms of quantum field theory and not about astrophysics, nuclear physics, or general relativity. Similarly, we try to avoid arguments based on thermodynamics, with the exception of some quantum statistics and the Boltzmann equation. With this in mind, these notes include material for at least 20 times 90 min of lectures for master-level students, preparing them for using the many excellent black-box tools which are available in the field. We really think that these notes only make sense if people print them out and go through the formulas one by one. This way any reader is bound to find a lot of typos, and we would be grateful if they could send us an email with them.

Finally, the literature listed at the end of the notes is not meant to cite original or relevant research papers. Instead, it gives examples for reviews or advanced lecture notes supplementing our lecture notes in different directions. Going through some of these mostly introductory papers will be instructive and fun once the basics have been covered by these lecture notes.

Durham, UK   Martin Bauer
Heidelberg, Germany   Tilman Plehn
November 2018

Acknowledgements

TP would like to thank many friends who have taught him dark matter, starting with the always-inspiring Dan Hooper. Dan also introduced him to deep-fried cheese curds and to the best ribs in Chicago. Tim Tait was of great help in at least two ways: for years he showed us that it is fun to work on dark matter even as a trained collider physicist, and then he answered every single email during the preparation of these notes. Our experimental co-lecturer Teresa Marrodan Undagoitia showed not only our students but also us how inspiring dark matter physics can be. As coauthors Joe Bramante, Adam Martin, and Paddy Fox gave us a great course on dark matter physics while we were writing these papers. Pedro Ruiz-Femenia was extremely helpful explaining the field theory behind the Sommerfeld enhancement to us. Jörg Jäckel for a long time and over many coffees tried to convince everybody that the axion is a great dark matter candidate. Teaching and discussing with Björn-Malte Schäfer was an excellent course on how thermodynamics is actually useful. Finally, there are many people who helped us with valuable advice while we prepared this course, like John Beacom, Martin Schmaltz, and Felix Kahlhöfer, and people who commented on the notes, like Elias Bernreuther, Johann Brehmer, Michael Baker, Anja Butter, Björn Eichmann, Ayres Freitas, Jan Horak, Michael Ratz, or Michael Schmidt.

Contents

Chapter 1
History of the Universe

When we study the history of the Universe with a focus on the matter content of the Universe, we have to define three key parameters:

- the Hubble constant H_0 which describes the expansion of the Universe. Two objects anywhere in the Universe move away from each other with a velocity proportional to their current distance r. The proportionality constant is defined through Hubble's law

$$H_0 := \frac{\dot{r}}{r} \approx 70 \, \frac{\text{km}}{\text{s Mpc}}$$

$$= 70 \, \frac{10^5 \, \text{cm}}{3.1 \cdot 10^{24} \, \text{cm}} \, \frac{1}{\text{s}} = 2.3 \cdot 10^{-18} \, 6.6 \cdot 10^{-16} \, \text{eV} = 1.5 \cdot 10^{-33} \, \text{eV}. \tag{1.1}$$

Throughout these lecture notes we will use these high-energy units with $\hbar = c = 1$, eventually adding $k_B = 1$. Because H_0 is not at all a number of order one we can replace H_0 with the dimensionless ratio

$$h := \frac{H_0}{100 \, \frac{\text{km}}{\text{s Mpc}}} \approx 0.7. \tag{1.2}$$

The Hubble 'constant' H_0 is defined at the current point in time, unless explicitly stated otherwise.

© Springer Nature Switzerland AG 2019
M. Bauer, T. Plehn, *Yet Another Introduction to Dark Matter*,
Lecture Notes in Physics 959, https://doi.org/10.1007/978-3-030-16234-4_1

– the cosmological constant Λ, which describes most of the energy content of the Universe and which is defined through the gravitational Einstein-Hilbert action

$$S_{EH} \equiv \frac{M_{Pl}^2}{2} \int d^4x \, \sqrt{-g} \, (R - 2\Lambda). \tag{1.3}$$

The reduced Planck mass is defined as

$$M_{Pl} = \frac{1}{\sqrt{8\pi G}} = 2.4 \cdot 10^{27} \, \text{eV}. \tag{1.4}$$

It is most convenient to also combine the Hubble constant and the cosmological constant to a dimensionless parameter

$$\boxed{\Omega_\Lambda := \frac{\Lambda}{3H_0^2}.} \tag{1.5}$$

– the matter content of the Universe which changes with time. As a mass density we can define it as ρ_m, but as for our other two key parameters we switch to the dimensionless parameters

$$\boxed{\Omega_m := \frac{\rho_m}{\rho_c}} \quad \text{and} \quad \Omega_r := \frac{\rho_r}{\rho_c}. \tag{1.6}$$

The denominator ρ_c is defined as the critical density separating an expanding from a collapsing Universe with $\Lambda = 0$. If we study the early Universe, we need to consider a sum of the relativistic matter or radiation content Ω_r and non-relativistic matter Ω_m alone. Today, we can also separate the non-relativistic baryonic matter content of the Universe. This is the matter content present in terms of atoms and molecules building stars, planets, and other astrophysical objects. The remaining matter content is dark matter, which we indicate with an index χ

$$\Omega_b := \frac{\rho_b}{\rho_c} \quad \Rightarrow \quad \boxed{\Omega_\chi := \Omega_m - \Omega_b.} \tag{1.7}$$

If the critical density ρ_c separates an expanding universe (described by H_0) and a collapsing universe (driven by the gravitational interaction G) we can guess that it should be given by something like a ratio of H_0 and G. Because the unit of the critical density has to be eV^4 we can already guess that $\rho_c \sim M_{Pl}^2 H_0^2$.

In classical gravity we can estimate ρ_c by computing the escape velocity of a massive particle outside a spherical piece of the Universe expanding according to Hubble's law. We start by computing the velocity v_{esc} a massive particle has to have to escape a gravitational field. Classically, it is defined by equal kinetic energy and gravitational binding energy for a test mass m at radius r,

$$\frac{m v_{esc}^2}{2} \overset{!}{=} \frac{GmM}{r} = \frac{Gm\frac{4\pi r^3}{3}\rho_c}{\frac{v_{esc}}{H_0}} = \frac{\frac{mr^3}{6M_{Pl}^2}\rho_c}{\frac{v_{esc}}{H_0}} \quad \text{with} \quad G = \frac{1}{8\pi M_{Pl}^2}$$

$$\Leftrightarrow H_0^3 r^3 \overset{\text{Eq.(1.1)}}{=} v_{esc}^3 \overset{!}{=} \frac{1}{3M_{Pl}^2}H_0\rho_c r^3$$

$$\Leftrightarrow \rho_c = 3M_{Pl}^2 H_0^2 = (2.5 \cdot 10^{-3} \text{ eV})^4 . \tag{1.8}$$

We give the numerical value based on the current Hubble expansion rate. For a more detailed account for the history of the Universe and a more solid derivation of ρ_c we will resort to the theory of general relativity in the next section.

1.1 Expanding Universe

Before we can work on dark matter as a major constituent of our observable Universe we need to derive a few key properties based on general relativity. For example, the Hubble constant H_0 as given by the linear relation in Eq. (1.1) is not actually a constant. To describe the history of the Universe through the time dependence of the Hubble constant $H(t)$ we start with the definition of a line element in flat space-time,

$$ds^2 = dt^2 - dr^2 - r^2 d\theta^2 - r^2 \sin^2 \theta d\phi^2$$

$$= \begin{pmatrix} dt \\ dr \\ rd\theta \\ r\sin\theta d\phi \end{pmatrix}^T \begin{pmatrix} 1 & 0 & 0 & 0 \\ 0 & -1 & 0 & 0 \\ 0 & 0 & -1 & 0 \\ 0 & 0 & 0 & -1 \end{pmatrix} \begin{pmatrix} dt \\ dr \\ rd\theta \\ r\sin\theta d\phi \end{pmatrix}. \tag{1.9}$$

The diagonal matrix defines the Minkowski metric, which we know from special relativity or from the covariant notation of electrodynamics. We can generalize this line element or metric to allow for a modified space-time, introducing a scale factor

a^2 as

$$ds^2 = dt^2 - \left(\frac{dr^2}{1 - \dfrac{r^2}{a^2}} + r^2 d\theta^2 + r^2 \sin^2 \theta d\phi^2 \right)$$

$$= dt^2 - a^2 \left(\frac{d\dfrac{r^2}{a^2}}{1 - \dfrac{r^2}{a^2}} + \frac{r^2}{a^2} d\theta^2 + \frac{r^2}{a^2} \sin^2 \theta d\phi^2 \right)$$

$$= dt^2 - a^2 \left(\frac{dr^2}{1 - r^2} + r^2 d\theta^2 + r^2 \sin^2 \theta d\phi^2 \right) \qquad \text{with} \quad \frac{r}{a} \to r. \qquad (1.10)$$

We define a to have the unit length or inverse energy and the last form of r to be dimensionless. In this derivation we implicitly assume a positive curvature through $a^2 > 0$. However, this does not have to be the case. We can allow for a free sign of the scale factor by introducing the free curvature k, with the possible values $k = -1, 0, 1$ for negatively, flat, or positively curved space. It enters the original form in Eq. (1.10) as

$$ds^2 = dt^2 - a^2 \left(\frac{dr^2}{1 - kr^2} + r^2 d\theta^2 + r^2 \sin^2 \theta d\phi^2 \right). \qquad (1.11)$$

At least for constant a this looks like a metric with a modified distance $r(t) \to r(t)a$. It is also clear that the choice $k = 0$ switches off the effect of $1/a^2$, because we can combine a and r to arrive at the original Minkowski metric.

Finally, there is really no reason to assume that the scale factor is constant with time. In general, the history of the Universe has to allow for a time-dependent scale factor $a(t)$, defining the line element or metric as

$$ds^2 = dt^2 - a(t)^2 \left(\frac{dr^2}{1 - kr^2} + r^2 d\theta^2 + r^2 \sin^2 \theta d\phi^2 \right). \qquad (1.12)$$

From Eq. (1.9) we can read off the corresponding metric including the scale factor,

$$g_{\mu\nu} = \begin{pmatrix} 1 & 0 & 0 & 0 \\ 0 & -\dfrac{a^2}{1 - kr^2} & 0 & 0 \\ 0 & 0 & -a^2 & 0 \\ 0 & 0 & 0 & -a^2 \end{pmatrix}. \qquad (1.13)$$

Now, the time-dependent scale factor $a(t)$ indicates a motion of objects in the Universe, $r(t) \to a(t)r(t)$. If we look at objects with no relative motion except for the expanding Universe, we can express Hubble's law given in Eq. (1.1) in terms of

$$r'(t) = a(t)\, r \quad \Leftrightarrow \quad \dot{r}'(t) = \dot{a}(t)\, r \overset{!}{=} H(t) r'(t) = H(t) a(t)\, r$$

$$\Leftrightarrow \quad \boxed{H(t) = \frac{\dot{a}(t)}{a(t)}}. \tag{1.14}$$

This relation reads like a linearized treatment of $a(t)$, because it depends only on the first derivative $\dot{a}(t)$. However, higher derivatives of $a(t)$ appear through a possible time dependence of the Hubble constant $H(t)$. From the above relation we can learn another, basic aspect of cosmology: we can describe the evolution of the universe in terms of

1. time t, which is fundamental, but hard to directly observe;
2. the Hubble constant $H(t)$ describing the expansion of the Universe;
3. the scale factor $a(t)$ entering the distance metric;
4. the temperature $T(t)$, which we will use from Sect. 1.2 on.

Which of these concepts we prefer depends on the kind of observations we want to link. Clearly, all of them should be interchangeable. For now we will continue with time.

Assuming the general metric of Eq. (1.12) we can solve Einstein's equation including the coupling to matter

$$R_{\mu\nu}(t) - \frac{1}{2} g_{\mu\nu}(t) R(t) + \Lambda(t) g_{\mu\nu}(t) = \frac{T_{\mu\nu}(t)}{M_{\text{Pl}}^2}. \tag{1.15}$$

The energy-momentum tensor includes the energy density $\rho_t = T_{00}$ and the corresponding pressure p. The latter is defined as the direction-independent contribution to the diagonal entries $T_{jj} = p_j$ of the energy-momentum tensor. The Ricci tensor $R_{\mu\nu}$ and Ricci scalar $R = g^{\mu\nu} R_{\mu\nu}$ are defined in terms of the metric; their explicit forms are one of the main topics of a lecture on general relativity. In terms of the scale factor the Ricci tensor reads

$$R_{00}(t) = -\frac{3\ddot{a}(t)}{a(t)} \quad \text{and} \quad R_{ij}(t) = \delta_{ij}\left(2\dot{a}(t)^2 + a(t)\ddot{a}(t)\right). \tag{1.16}$$

If we use the 00 component of Einstein's equation to determine the variable scale factor $a(t)$, we arrive at the Friedmann equation

$$\frac{\dot{a}(t)^2}{a(t)^2} + \frac{k}{a(t)^2} = \frac{\rho_t(t)}{3M_{\text{Pl}}^2} := \frac{\rho_m(t) + \rho_r(t) + \rho_\Lambda(t)}{3M_{\text{Pl}}^2} \quad \text{with}$$

$$\rho_\Lambda(t) := \Lambda(t) M_{\text{Pl}}^2 = 3H_0^2 M_{\text{Pl}}^2 \Omega_\Lambda(t), \tag{1.17}$$

with k defined in Eq. (1.11). A similar, second condition from the symmetry of the energy-momentum tensor and its derivatives reads

$$\frac{2\ddot{a}(t)}{a(t)} + \frac{\dot{a}(t)^2}{a(t)^2} + \frac{k}{a(t)^2} = -\frac{p(t)}{M_{\text{Pl}}^2}. \tag{1.18}$$

If we use the quasi-linear relation Eq. (1.14) and define the time-dependent critical total density of the Universe following Eq. (1.8), we can write the Friedmann equation as

$$H(t)^2 + \frac{k}{a(t)^2} = \frac{\rho_t(t)}{3M_{\text{Pl}}^2}$$

$$\Leftrightarrow \quad 1 + \frac{k}{H(t)^2 a(t)^2} = \frac{\rho_t(t)}{\rho_c(t)} =: \Omega_t(t) \qquad \text{with} \quad \boxed{\rho_c(t) := 3H(t)^2 M_{\text{Pl}}^2}. \tag{1.19}$$

This is the actual definition of the critical density $\rho_c(t)$. It means that k is determined by the time-dependent total energy density of the Universe,

$$k = H(t)^2 a(t)^2 \; (\Omega_t(t) - 1) \,. \tag{1.20}$$

This expression holds at all times t, including today, t_0. For $\Omega_t > 1$ the curvature is positive, $k > 0$, which means that the boundaries of the Universe are well defined. Below the critical density the curvature is negative. In passing we note that we can identify

$$\frac{\Lambda(t)}{3H(t)^2} \overset{\text{Eq. (1.5)}}{=} \Omega_\Lambda(t) \equiv \frac{\rho_\Lambda(t)}{\rho_c(t)} \overset{\text{Eq. (1.19)}}{=} \Lambda(t) M_{\text{Pl}}^2 \frac{1}{3H(t)^2 M_{\text{Pl}}^2}. \tag{1.21}$$

The two separate equations (Eqs. (1.17) and (1.18)) include not only the energy and matter densities, but also the pressure. Combining them we find

$$\frac{\ddot{a}(t)}{a(t)} \overset{\text{Eq. (1.18)}}{=} -\frac{1}{2}\left(\frac{\dot{a}(t)^2}{a(t)^2} + \frac{k}{a(t)^2}\right) - \frac{p(t)}{2M_{\text{Pl}}^2}$$

$$\overset{\text{Eq. (1.17)}}{=} -\frac{\rho_t(t)}{6M_{\text{Pl}}^2} - \frac{p(t)}{2M_{\text{Pl}}^2} = -\frac{\rho_t(t) + 3p(t)}{6M_{\text{Pl}}^2}. \tag{1.22}$$

The cosmological model based on Eq. (1.22) is called Friedmann–Lemaitre–Robertson–Walker model or FLRW model. In general, the relation between pressure

p and density ρ defines the thermodynamic equation of state

$$\boxed{p_j(t) = w_j \, \rho_j(t)} \qquad \text{with} \qquad w_j = \begin{cases} 0 & \text{non-relativistic matter} \\ 1/3 & \text{relativistic radiation} \\ -1 & \text{vacuum energy .} \end{cases}$$

$$(1.23)$$

It is crucial for our understanding of the matter content of the Universe. If we can measure w it will tell us what the energy or matter density of the Universe consists of.

Following the logic of describing the Universe in terms of the variable scale factor $a(t)$, we can replace the quasi-linear description in Eq. (1.14) with a full Taylor series for $a(t)$ around the current value a_0 and in terms of H_0. This will allow us to see the drastic effects of the different equations of state in Eq. (1.23),

$$a(t) - a_0 = \dot{a}(t_0)\,(t - t_0) + \frac{1}{2}\ddot{a}(t_0)\,(t - t_0)^2 + \mathcal{O}\left((t - t_0)^3\right)$$

$$\equiv a_0 H_0\,(t - t_0) - \frac{1}{2}a_0 q_0 H_0^2\,(t - t_0)^2 + \mathcal{O}\left((t - t_0)^3\right), \qquad (1.24)$$

implicitly defining q_0. The units are correct, because the Hubble constant defined in Eq. (1.1) is measured in energy. The pre-factors in the quadratic term are historic, as is the name deceleration parameter for q_0. Combined with our former results we find for the quadratic term

$$q_0 = -\frac{\ddot{a}(t_0)}{a_0 H_0^2} \overset{\text{Eq. (1.22)}}{=} \frac{\rho_t(t_0) + 3p(t_0)}{6H_0^2 M_{\text{Pl}}^2}$$

$$= \frac{1}{6H_0^2 M_{\text{Pl}}^2}\left(\rho_t(t_0) + 3\sum_j p_j(t_0)\right) \overset{\text{Eq. (1.23)}}{=} \frac{1}{2}\left(\Omega_t(t_0) + 3\sum_j \Omega_j(t_0)w_j\right).$$

$$(1.25)$$

The sum includes the three components contributing to the total energy density of the Universe, as listed in Eq. (1.31). Negative values of w corresponding to a Universe dominated by its vacuum energy can lead to negative values of q_0 and in turn to an accelerated expansion beyond the linear Hubble law. This is the basis for a fundamental feature in the evolution of the Universe, called inflation.

To be able to track the evolution of the Universe in terms of the scale factor $a(t)$ rather than time, we next compute the time dependence of $a(t)$. As a starting point, the Friedmann equation gives us a relation between $a(t)$ and $\rho(t)$. What we need is a relation of ρ and t, or alternatively a second relation between $a(t)$ and $\rho(t)$. Because we skip as much of general relativity as possible we leave it as an exercise to show that from the vanishing covariant derivative of the energy-momentum tensor, which gives rise to Eq. (1.18), we can also extract the time dependence of the energy and matter densities,

$$\frac{d}{dt}\left(\rho_j a^3\right) = -p_j \frac{d}{dt} a^3.$$

(1.26)

It relates the energy inside the volume a^3 to the work through the pressure p_j. From this conservation law we can extract the a-dependence of the energy and matter densities

$$\dot{\rho}_j a^3 + 3\rho_j a^2 \dot{a} = -3p_j a^2 \dot{a}$$

$$\Leftrightarrow \quad \dot{\rho}_j + 3\rho_j \left(1 + w_j\right)\frac{\dot{a}}{a} = 0$$

$$\Leftrightarrow \qquad\qquad \frac{\dot{\rho}_j}{\rho_j} = -3(1 + w_j)\frac{\dot{a}}{a}$$

$$\Leftrightarrow \qquad\qquad \log \rho_j = -3(1 + w_j) \log a + C$$

$$\Leftrightarrow \qquad \rho_j(a) = C\, a^{-3(1+w_j)} \propto \begin{cases} a^{-3} & \text{non-relativistic matter} \\ a^{-4} & \text{relativistic radiation} \\ \text{const} & \text{vacuum energy .} \end{cases}$$

(1.27)

This functional dependence is not yet what we want. To compute the time dependence of the scale factor $a(t)$ we use a power-law ansatz for $a(t)$ to find

$$\frac{\ddot{a}(t)}{a(t)} \overset{\text{Eq. (1.22)}}{=} -\frac{1+3w_j}{6M_{\text{Pl}}^2}\rho_j(t) \overset{\text{Eq. (1.27)}}{=} -\frac{1+3w_j}{6M_{\text{Pl}}^2}C\, a(t)^{-3(1+w_j)}$$

$$\Leftrightarrow \ddot{a}(t)a(t)^{2+3w_j} = \text{const}$$

$$\Leftrightarrow t^{\beta-2}\, t^{\beta(2+3w_j)} = \text{const} \quad \text{assuming } a \propto t^\beta$$

$$\Leftrightarrow t^{3\beta+3w_j\beta-2} = \text{const} \equiv t^0 \Leftrightarrow \beta = \frac{2}{3+3w_j}.$$

(1.28)

We can translate the result for $a(t) \propto t^\beta$ into the time-dependent Hubble constant

$$H(t) = \frac{\dot{a}(t)}{a(t)} \sim \frac{\beta\, t^{\beta-1}}{t^\beta} = \frac{\beta}{t} = \frac{2}{3+3w_j}\frac{1}{t}.$$

(1.29)

The problem with these formulas is that the power-law ansatz and the form of $H(t)$ obviously fails for the vacuum energy with $w = -1$. For an energy density only based on vacuum energy and neglecting any curvature, $k \equiv 0$, in the absence of matter, Eq. (1.14) together with the Friedmann equation becomes

$$H(t)^2 = \frac{\dot{a}(t)^2}{a(t)^2} \overset{\text{Eq. (1.17)}}{=} \frac{\rho_\Lambda(t)}{3M_{\text{Pl}}^2} = \frac{\Lambda(t)}{3}$$

$$\Leftrightarrow \quad a(t) = e^{H(t)t} = e^{\sqrt{\Lambda(t)/3}\,t}. \tag{1.30}$$

Combining this result and Eq. (1.28), the functional dependence of $a(t)$ reads

$$a(t) \sim \begin{cases} t^{2/(3+3w_j)} = \begin{cases} t^{2/3} & \text{non-relativistic matter} \\ t^{1/2} & \text{relativistic radiation} \\ & \end{cases} \\ e^{\sqrt{\Lambda(t)/3}\,t} & \text{vacuum energy.} \end{cases} \tag{1.31}$$

Alternatively, we can write for the Hubble parameter

$$H(t) \sim \begin{cases} \dfrac{2}{3+3w}\dfrac{1}{t} = \begin{cases} \dfrac{2}{3t} & \text{non-relativistic matter} \\ \dfrac{1}{2t} & \text{relativistic radiation} \\ & \end{cases} \\ \sqrt{\dfrac{\Lambda(t)}{3}} & \text{vacuum energy.} \end{cases} \tag{1.32}$$

From the above list we have now understood the relation between the time t, the scale factor $a(t)$, and the Hubble constant $H(t)$. An interesting aspect is that for the vacuum energy case $w = -1$ the change in the scale factor and with it the expansion of the Universe does not follow a power law, but an exponential law, defining an inflationary expansion. What is missing from our list at the beginning of this section is the temperature as the parameter describing the evolution of the Universe. Here we need to quote a thermodynamic result, namely that for constant entropy[1]

$$a(T) \propto \frac{1}{T}. \tag{1.33}$$

This relation is correct if the degrees of freedom describing the energy density of the Universe does not change. The easy reference point is $a_0 = 1$ today. We will use an improved scaling relation in Chap. 3.

Finally, we can combine several aspects described in these notes and talk about distance measures and their link to (i) the curved space-time metric, (ii) the expansion of the Universe, and (iii) the energy and matter densities. We will need it

---

[1]This is the only thermodynamic result which we will (repeatedly) use in these notes.

to discuss the cosmic microwave background in Sect. 1.4. As a first step, we compute the apparent distance along a line of sight, defined by $d\phi = 0 = d\theta$. This is the path of a traveling photon. Based on the time-dependent curved space-time metric of Eq. (1.12) we find

$$0 \overset{!}{=} ds^2 = dt^2 - a(t)^2 \frac{dr^2}{1 - kr^2} \qquad \Leftrightarrow \qquad dt = a(t) \frac{dr}{\sqrt{1 - kr^2}}. \qquad (1.34)$$

For the definition of the co-moving distance we integrate along this path,

$$\frac{d^c}{a_0} := \int \frac{dr}{\sqrt{1 - kr^2}} = \int \frac{dt}{a(t)} = \int da \frac{1}{\dot{a}(t)a(t)}. \qquad (1.35)$$

The distance measure we obtain from integrating dr in the presence of the curvature k is called the co-moving distance. It is the distance a photon traveling at the speed of light can reach in a given time. We can evaluate the integrand using the Friedmann equation, Eq. (1.17), and the relation $\rho\, a^{3(1-w)} = \text{const}$,

$$\dot{a}(t)^2 = a(t)^2 \frac{\rho_t(t)}{3M_{\text{Pl}}^2} - k$$

$$\overset{\text{Eq. (1.27)}}{=} \frac{\rho_m(t_0)a_0^3}{3M_{\text{Pl}}^2 a(t)} + \frac{\rho_r(t_0)a_0^4}{3M_{\text{Pl}}^2 a(t)^2} + \frac{\rho_\Lambda a(t)^2}{3M_{\text{Pl}}^2} - k$$

$$\overset{\text{Eq. (1.20)}}{=} H_0^2 \left[\Omega_m(t_0)\frac{a_0^3}{a(t)} + \Omega_r(t_0)\frac{a_0^4}{a(t)^2} \right.$$

$$\left. + \Omega_\Lambda a(t)^2 - (\Omega_t(t_0) - 1)\, a_0^2 \right]$$

$$\Rightarrow \quad \frac{1}{\dot{a}(t)a(t)}$$

$$= \frac{1}{H_0 \left[\Omega_m(t_0)a_0^3 a(t) + \Omega_r(t_0)a_0^4 + \Omega_\Lambda a(t)^4 - (\Omega_t(t_0) - 1)\, a_0^2 a(t)^2 \right]^{1/2}}.$$
$$(1.36)$$

For the integral defined in Eq. (1.35) this gives

$$\frac{d^c}{a_0} = \frac{1}{H_0} \int \frac{da}{\left[\Omega_m(t_0)a_0^3 a(t) + \Omega_r(t_0)a_0^4 + \Omega_\Lambda a(t)^4 - (\Omega_t(t_0) - 1)\, a_0^2 a(t)^2 \right]^{1/2}}$$

$$\approx \frac{1}{H_0} \int \frac{da}{\left[(\Omega_t(t_0) - \Omega_\Lambda)a_0^3 a(t) + \Omega_\Lambda a(t)^4 - (\Omega_t(t_0) - 1)\, a_0^2 a(t)^2 \right]^{1/2}}$$

$$= \frac{1}{H_0} \int \frac{da}{\left[\Omega_t(t_0)(a_0^3 a(t) - a_0^2 a(t)^2) - \Omega_\Lambda(a_0^3 a(t) - a(t)^4) + a_0^2 a(t)^2 \right]^{1/2}}.$$
$$(1.37)$$

Here we assume (and confirm later) that today $\Omega_r(t_0)$ can be neglected and hence $\Omega_t(t_0) = \Omega_m(t_0) + \Omega_\Lambda$. What is important to remember that looking back the variable scale factor is always $a(t) < a_0$. The integrand only depends on all mass and energy densities describing today's Universe, as well as today's Hubble constant. Note that the co-moving distance integrates the effect of time passing while we move along the light cone in Minkowski space. It would therefore be well suited for example to see which regions of the Universe can be causally connected.

Another distance measure based on Eq. (1.11) assumes the same line of sight $d\phi = 0 = d\theta$, but also a synchronized time at both end of the measurement, $dt = 0$. This defines a purely geometric, instantaneous distance of two points in space,

$$d\theta = d\phi = dt = 0 \quad \Rightarrow \quad ds(t) = -a(t)\frac{dr}{\sqrt{1 - kr^2}}$$

with $\quad k \overset{\text{Eq. (1.20)}}{=} H_0^2 a_0^2(\Omega_t(t_0) - 1)$

$$\Rightarrow \quad d_A^c(t) := \int_d^0 ds = -a(t)\int_d^0 \frac{dr}{\sqrt{1 - kr^2}} = \begin{cases} \dfrac{a(t)}{\sqrt{k}} \arcsin(\sqrt{k}\,d) & k > 0 \\[2mm] a(t)d & k = 0 \\[2mm] \dfrac{a(t)}{\sqrt{|k|}} \arcsinh(\sqrt{|k|}\,d) & k < 0. \end{cases}$$

$$(1.38)$$

This angular diameter distance is time dependent, but because it fixes the time at both ends we can use it for geometrical analyses. It depends on the assumed constant distance d, which can for example be identified with the co-moving distance $d \equiv d^c$. The curvature is again expressed in terms of today's energy density and Hubble constant.

1.2   Radiation and Matter

To understand the implications of the evolution of the Universe following Eq. (1.27), we can look at the composition of the Universe in terms of relativistic states (radiation), non-relativistic states (matter including dark matter), and a cosmological constant Λ. Figure 1.1 shows that at very large temperatures the Universe is dominated by relativistic states. When the variable scale factor a increases, the relativistic energy density drops like $1/a^4$. At the same time, the non-relativistic energy density drops like $1/a^3$. This means that as long as the relativistic energy density dominates, the relative fraction of matter increases linear in a. Radiation and matter contribute the same amount to the entire energy density around $a_{eq} = 3\cdot10^{-4}$, a period known as matter-radiation equality. The cosmological constant does not change, which means eventually it will dominate. This starts happening around now.

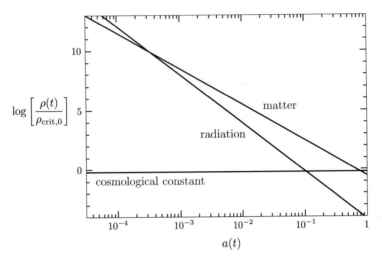

Fig. 1.1 Composition of our Universe as a function of the scale factor. Figure from Daniel Baumann's lecture notes [1]

We know experimentally that most of the matter content in the Universe is not baryonic, but dark matter. To describe its production in our expanding Universe we need to apply some basic statistical physics and thermodynamics. We start with the observation that according to Fig. 1.1 in the early Universe neither the curvature k nor the vacuum energy ρ_Λ play a role. This means that the relevant terms in the Friedmann equation Eq. (1.17) read

$$\boxed{H(t)^2 = \frac{\rho_m(t) + \rho_r(t)}{3M_{\mathrm{Pl}}^2}} \qquad \Leftrightarrow \qquad 1 = \frac{\rho_m(t) + \rho_r(t)}{\rho_c(t)} = \Omega_m(t) + \Omega_r(t).$$

$$(1.39)$$

This form will be the basis of our calculation in this section. The main change with respect to our above discussion will be a shift to temperature rather than time as an evolution variable.

For relativistic and non-relativistic particles or radiation we can use a unified picture in terms of their quantum fields. What we have to distinguish are fermion and boson fields and the temperature T relative to their respective masses m. The number of degrees of freedom are counted by a factor g, for example accounting for the anti-particle, the spin, or the color states. For example for the photon we have $g_\gamma = 2$, for the electron and positron $g_e = 2$ each, and for the left-handed neutrino $g_\nu = 1$. If we neglect the chemical potential because we assume to be either clearly non-relativistic or clearly relativistic, and we set $k_B = 1$, we (or better

MATHEMATICA) find

$$n_{eq}(T) = g \int \frac{d^3 p}{(2\pi)^3} \frac{1}{e^{E/T} \pm 1} \qquad \text{for fermions/bosons} \qquad (1.40)$$

$$= g \, 4\pi \int_m^\infty \frac{E \, dE}{(2\pi)^3} \frac{\sqrt{E^2 - m^2}}{e^{E/T} \pm 1} \qquad \text{using } E^2 = p^2 + m^2 \text{ and } p \, dp = E \, dE$$

$$= \begin{cases} g \left(\dfrac{mT}{2\pi}\right)^{3/2} e^{-m/T} & \text{non-relativistic states } T \ll m \\[2ex] \dfrac{\zeta_3}{\pi^2} \, g T^3 & \text{relativistic bosons } T \gg m \\[2ex] \dfrac{3}{4} \dfrac{\zeta_3}{\pi^2} \, g T^3 & \text{relativistic fermions } T \gg m. \end{cases}$$

The Riemann zeta function has the value $\zeta_3 = 1.2$. As expected, the quantum-statistical nature only matters once the states become relativistic and probe the relevant energy ranges. Similarly, we can compute the energy density in these different cases.

$$\rho_{eq}(T) = g \int \frac{d^3 p}{(2\pi)^3} \frac{E}{e^{E/T} \pm 1} = g \, 4\pi \int_m^\infty \frac{E \, dE}{(2\pi)^3} \frac{E\sqrt{E^2 - m^2}}{e^{E/T} \pm 1} \qquad (1.41)$$

$$= \begin{cases} mg \left(\dfrac{mT}{2\pi}\right)^{3/2} e^{-m/T} & \text{non-relativistic states } T \ll m \\[2ex] \dfrac{\pi^2}{30} \, g T^4 & \text{relativistic bosons } T \gg m \\[2ex] \dfrac{7}{8} \dfrac{\pi^2}{30} \, g T^4 & \text{relativistic fermions } T \gg m. \end{cases}$$

In the non-relativistic case the relative scaling of ρ relative to the number density is given by an additional factor $m \gg T$. In the relativistic case the additional factor is the temperature T, resulting in a Stefan–Boltzmann scaling of the energy density, $\rho \propto T^4$. To compute the pressure we can simply use the equation of state, Eq. (1.23), with $w = 1/3$.

The number of active degrees of freedom in our system depends on the temperature. As an example, above the electroweak scale $v = 246$ GeV the effective number of degrees of freedom includes all particles of the Standard Model

$$g_{\text{fermion}} = g_{\text{quark}} + g_{\text{lepton}} + g_{\text{neutrino}} = 6 \times 3 \times 2 \times 2 + 3 \times 2 \times 2 + 3 \times 2 = 90$$

$$g_{\text{boson}} = g_{\text{gluon}} + g_{\text{weak}} + g_{\text{photon}} + g_{\text{Higgs}} = 8 \times 2 + 3 \times 3 + 2 + 1 = 28.$$

$$(1.42)$$

Often, the additional factor $7/8$ for the fermions in Eq. (1.41) is absorbed in an effective number of degrees of freedom, implicitly defined through the unified relation

$$\rho_r = \frac{\pi^2}{30}\, g_{\text{eff}}(T)\, T^4\,,\tag{1.43}$$

with the relativistic contribution to the matter density defined in Eq. (1.17). Strictly speaking, this relation between the relativistic energy density and the temperature only holds if all states contributing to ρ_r have the same temperature, i.e. are in thermal equilibrium with each other. This does not have to be the case. To include different states with different temperatures we define g_{eff} as a weighted sum with the specific temperatures of each component, namely

$$g_{\text{eff}}(T) = \sum_{\text{bosons}} g_b \frac{T_b^4}{T^4} + \sum_{\text{fermions}} \frac{7}{8} g_f \frac{T_f^4}{T^4}\,.\tag{1.44}$$

For the entire Standard Model particle content at equal temperatures this gives

$$g_{\text{eff}}(T > 175\,\text{GeV}) \stackrel{\text{Eq. (1.42)}}{=} 28 + \frac{7}{8}\, 90 = 106.75.\tag{1.45}$$

When we reduce the temperature, this number of active degrees of freedom changes whenever a particle species vanishes at the respective threshold $T = m$. This curve

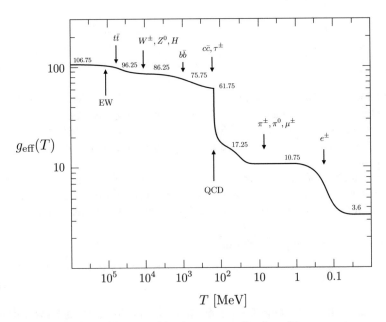

Fig. 1.2 Number of effective degrees of freedom g_{eff} as a function of the temperature, assuming the Standard Model particle content. Figure from Daniel Baumann's lecture notes [1]

is illustrated in Fig. 1.2. For today's value we will use the value

$$g_{\mathrm{eff}}(T_0) = 3.6. \tag{1.46}$$

Finally, we can insert the relativistic matter density given in Eq. (1.43) into the Friedmann equation Eq. (1.39) and find for the relativistic, radiation-dominated case

$$H(t)^2 \stackrel{\mathrm{Eq.\,(1.32)}}{=} \left(\frac{1}{2t}\right)^2 \stackrel{\mathrm{Eq.\,(1.39)}}{=} \frac{\rho_r}{3M_{\mathrm{Pl}}^2} \stackrel{\mathrm{Eq.\,(1.43)}}{=} \frac{1}{3M_{\mathrm{Pl}}^2} \frac{\pi^2}{30} g_{\mathrm{eff}}(T)\, T^4 = \left(\frac{\pi\sqrt{g_{\mathrm{eff}}}}{\sqrt{90}} \frac{T^2}{M_{\mathrm{Pl}}}\right)^2 . \tag{1.47}$$

This relation is important, because it links time, temperature, and Hubble constant as three possible scales in the evolution of our Universe in the relativistic regime. The one thing we need to check is if all relativistic relics have the same temperature.

1.3 Relic Photons

Before we will eventually focus on weakly interacting massive particles, forming the dark matter content of the Universe, it is for many reasons instructive to understand the current photon density. We already know that the densities of all particles pair-produced from a thermal bath in the early, hot Universe follows Eq. (1.41) and hence drops rapidly with the decreasing temperature of the expanding Universe. This kind of behavior is described by the Boltzmann equation, which we will study in some detail in Chap. 3. Computing the neutrino or photon number densities from the Boltzmann equation as a function of time or temperature will turn out to be a serious numerical problem. An alternative approach is to keep track of the relevant degrees of freedom $g(T)$ and compute for example the neutrino relic density ρ_ν from Eq. (1.43), all as a function of the temperature instead of time. In this approach it is crucial to know which particles are in equilibrium at any given point in time or temperature, which means that we need to track the temperature of the photon–neutrino–electron bath falling apart.

Neutrinos, photons, and electrons maintain thermal equilibrium through the scattering processes

$$\bar{\nu}_e\, e^- \to W^* \to \bar{\nu}_e\, e^- \qquad \text{and} \qquad e^-\, \gamma \to e^* \to e^-\, \gamma. \tag{1.48}$$

For low temperatures or energies $m_\nu \ll T, E \ll m_W$ the two cross sections are approximately

$$\sigma_{\nu e}(T) = \frac{\pi \alpha^2 T^2}{s_w^4 m_W^4} \ll \sigma_{\gamma e}(T) = \frac{\pi \alpha^2}{m_e^2}. \tag{1.49}$$

The coupling strength $g \equiv e/\sin\theta_w \equiv e/s_w$ with $s_w^2 \approx 1/4$ defines the weak coupling $\alpha = e^2/(4\pi) \approx 1/137$. The geometric factor π comes from the angular integration and helps us getting to the correct approximate numbers. The photons are more strongly coupled to the electron bath, which means they will decouple last, and in their decoupling we do not have to consider the neutrinos anymore. The interaction rate

$$\Gamma := \sigma\, v\, n \tag{1.50}$$

describes the probability for example of the neutrino or photon scattering process in Eq. (1.48) to happen. It is a combination of the cross section, the relevant number density and the velocity, measured in powers of temperature or energy, or inverse time. In our case, the relativistic relics move at the speed of light. Because the Universe expands, the density of neutrinos, photons, and charged leptons will at some point drop to a point where the processes in Eq. (1.48) hardly occur. They will stop maintaining the equilibrium between photons, neutrinos, and charged leptons roughly when the respective interaction rate drops below the Hubble expansion. This gives us the condition

$$\boxed{\frac{\Gamma(T_{\text{dec}})}{H(T_{\text{dec}})} \overset{!}{=} 1}. \tag{1.51}$$

as an implicit definition of the decoupling temperature.

Alternatively, we can compare the mean free path of the neutrinos or photons, $1/(\sigma\, n)$, to the Hubble length v/H to define the point of decoupling implicitly as

$$\frac{1}{\sigma(T_{\text{dec}})\, n(T_{\text{dec}})} \overset{!}{=} \frac{v}{H(T_{\text{dec}})} \qquad \Leftrightarrow \qquad \frac{\sigma(T_{\text{dec}})\, v\, n(T_{\text{dec}})}{H(T_{\text{dec}})} \overset{!}{=} 1. \tag{1.52}$$

While the interaction rate for example for neutrino–electron scattering is in the literature often defined using the neutrino density $n = n_\nu$. For the mean free path we have to use the target density, in this case the electron $n = n_e$.

We should be able to compute the photon decoupling from the electrons based on the above definition of T_{dec} and the photon–electron or Thomson scattering rate in Eq. (1.49). The problem is, that it will turn out that at the time of photon decoupling the electrons are no longer the relevant states. Between temperatures of 1 MeV and the relevant eV-scale for photon decoupling, nucleosynthesis will have happened, and the early Universe will be made up by atoms and photons, with a small number of free electrons. Based on this, we can very roughly guess the temperature at which the Universe becomes transparent to photons from the fact that most of the electrons are bound in hydrogen atoms. The ionization energy of hydrogen is 13.6 eV, which is our first guess for T_{dec}. On the other hand, the photon temperature will follow a Boltzmann distribution. This means that for a given temperature T_{dec} there will be a high-energy tail of photons with much larger energies. To avoid having too many photons still ionizing the hydrogen atoms the photon temperature should therefore come out as $T_{\text{dec}} \lesssim 13.6$ eV.

Going back to the defining relation in Eq. (1.51), we can circumvent the problem of the unknown electron density by expressing the density of free electrons first relative to the density of electrons bound in mostly hydrogen, with a measured suppression factor $n_e/n_B \approx 10^{-2}$. Moreover, we can relate the full electron density or the baryon density n_B to the photon density n_γ through the measured baryon–to–photon ratio. In combination, this gives us for the time of photon decoupling

$$n_e(T_{\text{dec}}) = \frac{n_e}{n_B}(T_{\text{dec}}) \, n_B(T_{\text{dec}})$$

$$= \frac{n_e}{n_B}(T_{\text{dec}}) \frac{n_B}{n_\gamma}(T_{\text{dec}}) \, n_\gamma(T_{\text{dec}}) = 10^{-2} \, 10^{-10} \frac{2\zeta_3 T_{\text{dec}}^3}{\pi^2}. \qquad (1.53)$$

At this point we only consider the ratio $n_B/n_\gamma \approx 10^{-10}$ a measurable quantity, its meaning will be the topic of Sect. 2.4. With this estimate of the relevant electron density we can compute the temperature at the point of photon decoupling. For the Hubble constant we need the number of active degrees of freedom in the absence of neutrinos and just including electrons, positions, and photons

$$g_{\text{eff}}(T_{\text{dec}}) = \frac{7}{8}(2+2) + 2 = 5.5. \qquad (1.54)$$

Inserting the Hubble constant from Eq. (1.47) and the cross section from Eq. (1.49) gives us the condition

$$\frac{\Gamma_\gamma}{H} = \frac{2\pi\zeta_3\alpha^2}{\pi^2} 10^{-12} \frac{T^3}{m_e^2} \frac{\sqrt{90}M_{\text{Pl}}}{\pi} \frac{1}{\sqrt{g_{\text{eff}}(T)}T^2}$$

$$= \frac{6\sqrt{10}\,\zeta_3}{\pi^2} 10^{-12}\alpha^2 \frac{1}{\sqrt{g_{\text{eff}}(T)}} \frac{M_{\text{Pl}}T}{m_e^2} \overset{!}{=} 1$$

$$\Leftrightarrow \quad T_{\text{dec}} = 10^{12} \frac{\pi^2}{6\sqrt{10}\,\zeta_3} \frac{m_e^2}{M_{\text{Pl}}} \frac{\sqrt{g_{\text{eff}}(T_{\text{dec}})}}{\alpha^2} \approx (0.1 \dots 1) \, \text{eV}. \qquad (1.55)$$

As discussed above, to avoid having too many photons still ionizing the hydrogen atoms, the photon temperature indeed is $T_{\text{dec}} \approx 0.26 \, \text{eV} < 13.6 \, \text{eV}$.

These decoupled photons form the cosmic microwave background (CMB), which will be the main topic of Sect. 1.4. The main property of this photon background, which we will need all over these notes, is its current temperature. We can compute $T_{0,\gamma}$ from the temperature at the point of decoupling, when we account for the expansion of the Universe between T_{dec} and now. We can for example use the time evolution of the Hubble constant $H \propto T^2$ from Eq. (1.47) to compute the photon temperature today. We find the experimentally measured value of

$$T_{0,\gamma} = 2.4 \cdot 10^{-4} \, \text{eV} = 2.73 \, \text{K} \approx \frac{T_{\text{dec}}}{1000}. \qquad (1.56)$$

This energy corresponds to a photon frequency around 60 GHz, which is in the microwave range and inspires the name CMB. We can translate the temperature at the time of photon decoupling into the corresponding scale factor,

$$a_{dec} \overset{\text{Eq.}(1.33)}{=} a_0 \frac{T_{0,\gamma}}{T_{dec}} = \frac{T_{0,\gamma}}{T_{dec}} = \frac{2.4 \cdot 10^{-4}\,\text{eV}}{0.26\,\text{eV}} \approx \frac{1}{1100}. \tag{1.57}$$

From Eq. (1.40) we can also compute the current density of CMB photons,

$$n_\gamma(T_0) = \frac{2\zeta_3}{\pi^2}\, T_{0,\gamma}^3 = \frac{410}{\text{cm}^3}. \tag{1.58}$$

1.4 Cosmic Microwave Background

In Sect. 1.3 we have learned that at temperatures around 0.1 eV the thermal photons decoupled from the matter in the Universe and have since then been streaming through the expanding Universe. This is why their temperature has dropped to $T_0 = 2.4 \cdot 10^{-4}$ eV now. We can think of the cosmic microwave background or CMB photons as coming from a sphere of last scattering with the observer in the center. The photons stream freely through the Universe, which means they come from this sphere straight to us.

The largest effect leading to a temperature fluctuation in the CMB photons is that the earth moves through the photon background or any other background at constant speed. We can subtract the corresponding dipole correlation, because it does not tell us anything about fundamental cosmological parameters. The most important, fundamental result is that after subtracting this dipole contribution the temperature on the surface of last scattering only shows tiny variations around $\delta T/T \lesssim 10^{-5}$. The entire surface, rapidly moving away from us, should not be causally connected, so what generated such a constant temperature? Our favorite explanation for this is a phase of very rapid, inflationary period of expansion. This means that we postulate a fast enough expansion of the Universe, such that the sphere or last scattering becomes causally connected. From Eq. (1.31) we know that such an expansion will be driven not by matter but by a cosmological constant. The detailed structure of the CMB background should therefore be a direct and powerful probe for essentially all parameters defined and discussed in this chapter.

The main observable which the photon background offers is their temperature or energy—additional information for example about the polarization of the photons is very interesting in general, but less important for dark matter studies. Any effect which modifies this picture of an entirely homogeneous Universe made out of a thermal bath of electrons, photons, neutrinos, and possibly dark matter particles, should be visible as a modification to a constant temperature over the sphere of last scattering. This means, we are interested in analyzing temperature fluctuations between points on this surface.

The appropriate observables describing a sphere are the angles θ and ϕ. Moreover, we know that spherical harmonics are a convenient set of orthogonal basis functions which describe for example temperature variations on a sphere,

$$\frac{\delta T(\theta, \phi)}{T_0} := \frac{T(\theta, \phi) - T_0}{T_0} = \sum_{\ell=0}^{\infty} \sum_{m=-\ell}^{\ell} a_{\ell m} Y_{\ell m}(\theta, \phi). \tag{1.59}$$

The spherical harmonics are orthonormal, which means in terms of the integral over the full angle $d\Omega = d\phi d\cos\theta$

$$\int d\Omega\, Y_{\ell m}(\theta, \phi)\, Y_{\ell' m'}^*(\theta, \phi) = \delta_{\ell\ell'}\delta_{mm'}$$

$$\Rightarrow \qquad \int d\Omega\, \frac{\delta T(\theta, \phi)}{T_0} Y_{\ell' m'}^*(\theta, \phi) \stackrel{\text{Eq.}(1.59)}{=} \sum_{\ell m} a_{\ell m} \int d\Omega\, Y_{\ell m}(\theta, \phi)\, Y_{\ell' m'}^*(\theta, \phi)$$

$$= \sum_{\ell m} a_{\ell m} \delta_{\ell\ell'}\delta_{mm'} = a_{\ell' m'}. \tag{1.60}$$

This is the inverse relation to Eq. (1.59), which allows us to compute the set of numbers $a_{\ell m}$ from a known temperature map $\delta T(\theta, \phi)/T_0$.

For the function $T(\theta, \phi)$ measured over the sphere of last scattering, we can ask the three questions which we usually ask for distributions which we know are peaked:

1. what is the peak value?
2. what is the width of the peak?
3. what the shape of the peak?

For the CMB we assume that we already know the peak value T_0 and that there is no valuable information in the statistical distribution. This means that we can focus on the width or the variance of the temperature distribution. Its square root defines the standard deviation. In terms of the spherical harmonics the variance reads

$$\frac{1}{4\pi} \int d\Omega\, \left(\frac{\delta T(\theta, \phi)}{T_0}\right)^2 = \frac{1}{4\pi} \int d\Omega\, \left[\sum_{\ell m} a_{\ell m} Y_{\ell m}(\theta, \phi)\right]$$

$$\times \left[\sum_{\ell' m'} a_{\ell' m'}^* Y_{\ell' m'}^*(\theta, \phi)\right]$$

$$\stackrel{\text{Eq.}(1.60)}{=} \frac{1}{4\pi} \sum_{\ell m, \ell', m'} a_{\ell m} a_{\ell' m'}^* \delta_{\ell\ell'}\delta_{mm'} = \frac{1}{4\pi} \sum_{\ell m} |a_{\ell m}|^2. \tag{1.61}$$

We can further simplify this relation by our expectation for the distribution of the temperature deviations. We remember for example from quantum mechanics that for the angular momentum the index m describes the angular momentum in one specific direction. Our analysis of the surface of last scattering, just like the hydrogen atom without an external magnetic field, does not have any special direction. This implies that the values of $a_{\ell m}$ do not depend on the value of the index m; the sum over m should just become a sum over $2\ell + 1$ identical terms. We therefore define the observed power spectrum as the average of the $|a_{\ell m}|^2$ over m,

$$C_\ell := \frac{1}{2\ell + 1} \sum_{m=-\ell}^{\ell} |a_{\ell m}|^2$$

$$\Leftrightarrow \qquad \boxed{\frac{1}{4\pi} \int d\Omega \left(\frac{\delta T(\theta, \phi)}{T_0}\right)^2 = \sum_{\ell=0}^{\infty} \frac{2\ell + 1}{4\pi} C_\ell}. \tag{1.62}$$

The great simplification of this last assumption is that we now just analyze the discrete values C_ℓ as a function of $\ell \geq 0$.

Note that we analyze the fluctuations averaged over the surface of last scattering, which gives us one curve C_ℓ for discrete values $\ell \geq 0$. This curve is one measurement, which means none of its points have to perfectly agree with the theoretical expectations. However, because of the averaging over m possible statistical fluctuation will cancel, in particular for larger values of ℓ, where we average over more independent orientations.

We can compare the series in spherical harmonics Eq. (1.59) to a Fourier series. The latter will, for example, analyze the frequencies contributing to a sound from a musical instrument. The discrete series of Fourier coefficients tells us which frequency modes contribute how strongly to the sound or noise. The spherical harmonic do something similar, which we can illustrate using the properties of the $Y_{\ell 0}(\theta, \phi)$. Their explicit form in terms of the associated Legendre polynomials $P_{\ell m}$ and the Legendre polynomials P_ℓ is

$$Y_{\ell m}(\theta, \phi) = (-1)^m e^{im\phi} \sqrt{\frac{2\ell + 1}{4} \frac{(\ell - m)!}{(\ell + m)!}} \, P_{\ell m}(\cos\theta)$$

$$= (-1)^m e^{im\phi} \sqrt{\frac{2\ell + 1}{4} \frac{(\ell - m)!}{(\ell + m)!}}$$

$$\times (-1)^m \left(1 - \cos^2\theta\right)^{m/2} \frac{d^m}{d(\cos\theta)^m} \, P_\ell(\cos\theta)$$

$$\Rightarrow \qquad Y_{\ell 0}(\theta, \phi) = \frac{\sqrt{2\ell + 1}}{2} \, P_\ell(\cos\theta). \tag{1.63}$$

The Legendre polynomial is for example defined through

$$P_\ell(\cos\theta) = \frac{1}{2^\ell \ell!} \frac{d^\ell}{dt^\ell} (\cos^2\theta - 1)^\ell = C\cos^\ell\theta + \cdots, \qquad (1.64)$$

with the normalization $P_\ell(\pm 1) = 1$ and ℓ zeros in between. Approximately, these zeros occur at

$$P_\ell(\cos\theta) = 0 \qquad \Leftrightarrow \qquad \cos\theta = \cos\left(\pi\frac{4k-1}{4\ell+2}\right) \qquad k = 1,\ldots,\ell. \qquad (1.65)$$

The first zero of each mode defines an angular resolution θ_ℓ of the ℓth term in the hypergeometric series,

$$\cos\left(\pi\frac{3}{4\ell+2}\right) \equiv \cos\theta_\ell \qquad \Leftrightarrow \qquad \theta_\ell \approx \frac{3\pi}{4\ell}. \qquad (1.66)$$

This separation in angle can obviously be translated into a spatial distance on the sphere of last scattering, if we know the distance of the sphere of last scattering to us. This means that the series of $a_{\ell m}$ or the power spectrum C_ℓ gives us information about the angular distances (encoded in ℓ) which contribute to the temperature fluctuations $\delta T/T_0$.

Next, we need to think about how a distribution of the C_ℓ will typically look. In Fig. 1.3 we see that the measured power spectrum essentially consists of a set of peaks. Each peak gives us an angular scale with a particularly large

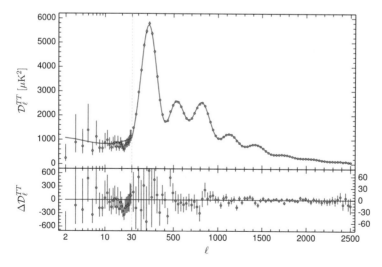

Fig. 1.3 Power spectrum as measured by PLANCK in 2015. Figure from the PLANCK collaboration [2]

contribution to the temperature fluctuations. The leading physics effects generating such temperature fluctuations are:

- acoustic oscillations which occur in the baryon–photon fluid at the time of photon decoupling. As discussed in this chapter the photons are initially strongly coupled to the still separate electrons and baryons, because the two components interact electromagnetically through Thomson scattering. Following Eq. (1.49) the weak interaction can be neglected in comparison to Thomson scattering for ordinary matter. On the other hand, we can see what happens when a sizeable fraction of the matter in the Universe is not baryonic and only interacts gravitationally and possibly through the weak interaction. Such new, dark matter generates gravitational wells around regions of large matter accumulation.

  The baryon–photon fluid gets pulled into these gravitational wells. For the relativistic photon gas we can relate the pressure to the volume and the temperature through the thermodynamic equation of state $PV \propto T$. If the temperature cannot adjust rapidly enough, for example in an adiabatic transition, a reduced volume will induce an increased pressure. This photon pressure acts against the gravitational well. The photons moving with and against a slope in the gravitational potential induces a temperature fluctuation located around regions of dark matter concentration. Such an oscillation will give rise to a tower of modes with definite wave lengths. For a classical box-shaped potential they will be equi-distant, while for a smoother potential the higher modes will be pulled apart. Strictly speaking, we can separate the acoustic oscillations into a temperature effect and a Doppler shift, which have separate effects on the CMB power spectrum.
- the effect of general relativity on the CMB photons, not only related to the decoupling, but also related to the propagation of the streaming photons to us. In general, the so-called Sachs–Wolfe effect describes this impact of gravity on the CMB photons. Such an effect occurs if large accumulations of mass or energy generate a distinctive gravitational potential which changes during the time the photons travel through it. This effect will happen before and while the photons are decoupling, but also during the time they are traveling towards us. From the discussion above it is clear that it is hard to separate the Sachs–Wolfe effect during photon decoupling from the other effects generating the acoustic oscillations. For the streaming photons we need to integrate the effect over the line of sight. The later the photons see such a gravitational potential, the more likely they are to probe the cosmological constant or the geometrical shape of the Universe close to today.

Figure 1.3 confirms that the power spectrum essentially consists of a set of peaks, i.e. a set of angular scales at which we observe a particularly strong correlation in temperatures. They are generated through the acoustic oscillations. Before we discuss the properties of the peaks we notice two general features: first, small values of ℓ lead to large error bars. This is because for large angular separations there are not many independent measurements we can do over the sphere, i.e. we lose the statistical advantage from combining measurements over the whole sphere in one C_ℓ curve. Second, the peaks are washed out for large ℓ. This happens because our approximation that the sphere of last scattering has negligible thickness catches up with us. If we take into account that the sphere of last scattering has a finite thickness, the strongly peaked structure of the power spectrum gets washed out. Towards large values ℓ or small distances the thickness effects become comparable to the spatial resolution at the time of last scattering. This leads to an additional damping term

$$C_\ell \propto e^{-\ell^2/1500^2},$$

(1.67)

which washes out the peaks above $\ell = 1500$ and erases all relevant information.

Next, we can derive the position of the acoustic peaks. Because of the rapid expansion of the Universe, a critical angle θ_ℓ in Eq. (1.66) defines the size of patches of the sky, which were not in causal contact during and since the time of the last scattering. Below the corresponding ℓ-value there will be no correlation. It is given by two distances: the first of them is the distance on the sphere of last scattering, which we can compute in analogy to the co-moving distance defined in Eq. (1.37). Because the co-moving distance is best described by an integral over the scale factor a, we use the value $a_{\text{dec}} \approx 1/1100$ from Eq. (1.57) to integrate the ratio of the distance to the sound velocity in the baryon–photon fluid c_s to

$$\frac{r_s}{c_s} \overset{\text{Eq. (1.35)}}{=} \int_0^{a_{\text{dec}}} \frac{da}{a(t)\dot{a}(t)}.$$

(1.68)

For a perfect relativistic fluid the speed of sound is given by $c_s = 1/\sqrt{3}$. This distance is called the sound horizon and depends mostly on the matter density around the oscillating baryon–photon fluid. The second relevant distance is the distance between us and the sphere of last scattering. Again, we start from the co-moving distance d^c introduced in Eq. (1.35). Following Eq. (1.37) it will depend on the current energy and matter content of the universe. The angular separation is

$$\sin \theta_\ell = \frac{r_s(\Omega_m, \Omega_b)}{d^c}.$$

(1.69)

Both $r_s(\Omega_m, \Omega_b)$ and d^c are described by the same integrand in Eq. (1.36). It can be simplified for a matter-dominated ($\Omega_r \ll \Omega_m$) and almost flat ($\Omega_t \approx \Omega_m$) Universe to

$$\left[\Omega_m(t_0)a(t) + \Omega_r(t_0) + \Omega_\Lambda a(t)^4 - (\Omega_t(t_0) - 1) a(t)^2\right]^{-1/2} \approx \frac{1}{\sqrt{\Omega_m(t_0)a(t)}},$$

(1.70)

where we also replaced $a_0 = 1$. The ratio of the two integrals then gives

$$
\sin \theta_\ell = \frac{c_s \displaystyle\int_0^{a_{\rm dec}} \frac{da}{\sqrt{a}}}{\displaystyle\int_{a_{\rm dec}}^1 \frac{da}{\sqrt{a}}} = \frac{1}{\sqrt{3}} \frac{\sqrt{a_{\rm dec}}}{1 - \sqrt{a_{\rm dec}}} \approx \frac{1}{55} \qquad \Rightarrow \qquad \theta_\ell \approx 1^\circ . \tag{1.71}
$$

A more careful calculation taking into account the reduced speed of sound and the effects from Ω_r, Ω_Λ gives a critical angle

$$
\theta_\ell \approx 0.6^\circ \qquad \overset{\text{Eq. (1.66)}}{\Rightarrow} \qquad \ell \Big|_{\text{first peak}} = \frac{4}{3\pi} 0.6^\circ = 225. \tag{1.72}
$$

The first peak in Fig. 1.3 corresponds to the fundamental tone, a sound wave with a wavelength twice the size of the horizon at decoupling. By the time of the last scattering this wave had just compressed once. Note that a closed or open universe predict different result for θ_ℓ following Eq. (1.38). The measurement of the position of the first peak is therefore considered a measurement of the geometry of the universe and a confirmation of its flatness.

The second peak corresponds to the sound wave which underwent one compression and one rarefaction at the time of the last scattering and so forth for the higher peaks. Even-numbered peaks are associated with how far the baryon–photon fluid compresses due to the gravitational potential, odd-numbered peaks indicate the rarefaction counter effect of radiative pressure. If the relative baryon content in the baryon–photon is higher, the radiation pressure decreases and the compression peaks become higher. The relative amplitude between odd and even peaks can therefore be used as a measure of Ω_b.

Dark matter does not respond to radiation pressure, but contributes to the gravitational wells and therefore further enhances the compression peaks with respect to the rarefaction peaks. This makes a large third peak a sign of a sizable dark matter component at the time of the last scattering.

From Fig. 1.1 we know that today we can neglect $\Omega_r(t_0) \ll \Omega_m(t_0) \sim \Omega_\Lambda$. Moreover, the relativistic matter content is known from the accurate measurement of the photon temperature T_0, giving $\Omega_r h^2$ through Eq. (2.9). This means that the peaks in the CMB power spectrum will be described by: the cosmological constant defined in Eq. (1.5), the entire matter density defined in Eq. (1.6), which is dominated by the dark matter contribution, as well as by the baryonic matter density defined in Eq. (1.7), and the Hubble parameter defined in Eq. (1.1). People usually choose the four parameters

$$
\boxed{\Omega_t(t_0), \quad \Omega_\Lambda, \quad \Omega_m(t_0)h^2, \quad \Omega_b(t_0)h^2} . \tag{1.73}
$$

Including h^2 in the matter densities means that we define the total energy density $\Omega_t(t_0)$ as an independent parameter, but at the expense of h or H_0 now being a derived quantity,

$$\left(\frac{H_0}{100\frac{\text{km}}{\text{s Mpc}}}\right)^2 = h^2 = \frac{\Omega_m(t_0)h^2}{\Omega_m(t_0)} = \frac{\Omega_m(t_0)h^2}{\Omega_t(t_0) - \Omega_\Lambda - \Omega_r(t_0)} \approx \frac{\Omega_m(t_0)h^2}{\Omega_t(t_0) - \Omega_\Lambda}.$$

(1.74)

There are other, cosmological parameters which we for example need to determine the distance of the sphere of last scattering, but we will not discuss them in detail. Obviously, the choice of parameter basis is not unique, but a matter of convenience. There exist plenty of additional parameters which affect the CMB power spectrum, but they are not as interesting for non-relativistic dark matter studies.

We go through the impact of the parameters basis defined in Eq. (1.73) one by one:

- Ω_t affects the co-moving distance, Eq. (1.37), such that an increase in $\Omega_t(t_0)$ decreases d^c. The same link to the curvature, $k \propto (\Omega_t(t_0) - 1)$ as given in Eq. (1.20), also decreases ds, following Eq. (1.38); this way the angular diameter distance d_A^c is reduced. In addition, there is an indirect effect through H_0; following Eq. (1.74) an increased total energy density decreases H_0 and in turn increases d^c.
  Combining all of these effects, it turns out that increasing $\Omega_t(t_0)$ decreases d^c. According to Eq. (1.69) a smaller predicted value of d^c effectively increases the corresponding θ_ℓ scale. This means that the acoustic peak positions consistently appear at smaller ℓ values.
- Ω_Λ has two effects on the peak positions: first, Ω_Λ enters the formula for d^c with a different sign, which means an increase in Ω_Λ also increases d^c and with it d^c. At the same time, an increased Ω_Λ also increases H_0 and this way decreases d^c. The combined effect is that an increase in Ω_Λ moves the acoustic peaks to smaller ℓ. Because in our parameter basis both, $\Omega_t(t_0)$ and Ω_Λ have to be determined by the peak positions we will need to find a way to break this degeneracy.
- $\Omega_m h^2$ is dominated by dark matter and provides the gravitational potential for the acoustic oscillations. Increasing the amount of dark matter stabilizes the gravitational background for the baryon–photon fluid, reducing the height of all peaks, most visibly the first two. In addition, an increased dark matter density makes the gravitational potential more similar to a box shape, bringing the higher modes closer together.
- $\Omega_b h^2$ essentially only affects the height of the peaks. The baryons provide most of the mass of the baryon–photon fluid, which until now we assume to be infinitely strongly coupled. Effects of a changed $\Omega_b h^2$ on the CMB power spectrum arise when we go beyond this infinitely strong coupling. Moreover, an increased amount of baryonic matter increases the height of the odd peaks and reduces the height of the even peaks.

Separating these four effects from each other and from other astrophysical and cosmological parameters obviously becomes easier when we can include more and higher peaks. Historically, the WMAP experiment lost sensitivity around the third peak. This means that its results were typically combined with other experiments. The PLANCK satellite clearly identified seven peaks and measures in a slight modification to our basis in Eq. (1.73) [2]

$$\boxed{\Omega_\chi h^2 = 0.1198 \pm 0.0015}$$

$$\Omega_b h^2 = 0.02225 \pm 0.00016$$

$$\Omega_\Lambda = 0.6844 \pm 0.0091$$

$$H_0 = 67.27 \pm 0.66 \, \frac{\text{km}}{\text{Mpc s}}. \tag{1.75}$$

The dark matter relic density is defined in Eq. (1.7). This is the best measurement of Ω_χ we currently have.

1.5   Structure Formation

A powerful tool to analyze the evolution of the Universe is the distribution of structures at different length scales, from galaxies to the largest structures. These structures are due to small primordial inhomogeneities, tiny gravitational wells disrupting the homogeneous and isotropic universe we have considered so far. They have then been amplified to produce the galaxies, galaxy groups and super-clusters we observe today. The leading theory for the origin of these perturbations is based on quantum fluctuations of in the inflation field, which is responsible for the epoch of exponential expansion of the universe. We leave the details of this idea to a cosmology lecture, but note that quantum fluctuations behave random or Gaussian. The evolution of these primordial seeds of over-densities with the expansion of the universe will give us information on the dark matter density and on dark matter properties.

We start with the evolution of a general matter density in the Universe in the presence of a gravitational field. As long as the cosmic structures are small compared to the curvature of the universe and we are not interested in the (potentially) relativistic motion of particles we can compute the evolution of density perturbations using Newtonian physics. The matter density ρ, the matter velocity $\vec{u}$, and its gravitational potential ϕ satisfy the equations

$$\frac{\partial \rho_m}{\partial t} = -\nabla \cdot (\rho_m \vec{u}) \qquad \text{continuity equation} \qquad (1.76)$$

$$\left(\frac{\partial}{\partial t} + \vec{u} \cdot \nabla\right) \vec{u} = -\frac{\nabla p}{\rho_m} - \nabla \phi \qquad \text{Euler equation} \qquad (1.77)$$

$$\nabla^2 \phi = 4\pi G \rho_m \qquad \text{Poisson equation,} \qquad (1.78)$$

where p denotes an additional pressure and $G = 1/(8\pi M_{\text{Pl}}^2)$ is the gravitational coupling defined in Eq. (1.4). This set of equations can be solved by a homogeneously expanding fluid

$$\rho = \rho(t_0) \left(\frac{a_0}{a}\right)^3 \qquad \vec{u} = \frac{\dot{a}}{a} \vec{r} = H\vec{r} \qquad \phi = \frac{1}{12M_{\text{Pl}}^2} \rho r^2 \qquad \nabla p = 0. \quad (1.79)$$

It is the Newtonian version of the matter-dominated Friedmann model. The Euler equation turns into the second Friedmann equation, Eq. (1.22), for a flat universe,

$$\dot{H}\vec{r} + H\vec{r} \cdot \nabla(H\vec{r}) = -\nabla\phi = -\frac{1}{6M_{\text{Pl}}^2} \rho_m \vec{r}$$

$$\Leftrightarrow \qquad \dot{H} + H^2 = -\frac{\rho_m}{6M_{\text{Pl}}^2} . \qquad (1.80)$$

The first Friedmann equation in this approximation also follows when we use Eq. (1.19) for $k \to 0$ and $\rho_t = \rho_m$,

$$H^2 = \frac{\rho_m}{3M_{\text{Pl}}^2} . \qquad (1.81)$$

We will now allow for small perturbations around the background given in Eq. (1.79),

$$\rho(t, \vec{r}) = \bar{\rho}(t) + \delta_\rho(t, \vec{r}) \qquad\qquad \vec{u}(t, \vec{r}) = H(t)\vec{r} + \vec{\delta}_u(t, \vec{r})$$

$$\phi(t, \vec{r}) = \frac{1}{12M_{\text{Pl}}^2} \bar{\rho} r^2 + \delta_\phi(t, \vec{r}) \qquad p(t, \vec{r}) = \bar{p}(t) + \delta_p(t, \vec{r}).$$

$$(1.82)$$

The pressure and density fluctuations are linked by the speed of sound $\delta_p = c_s^2 \delta_\rho$. Inserting Eq. (1.82), the continuity equation becomes

$$0 = \dot{\rho} + \nabla \cdot (\rho \vec{u})$$

$$= \dot{\bar{\rho}} + \dot{\delta}_\rho + \bar{\rho}\nabla \cdot (H\vec{r} + \vec{\delta}_u) + \nabla\delta_\rho \cdot (H\vec{r} + \vec{\delta}_u)$$

$$= \dot{\bar{\rho}} + \dot{\delta}_\rho + \bar{\rho} \nabla \cdot (H\vec{r} + \vec{\delta}_u) + \nabla \delta_\rho \cdot (H\vec{r} + \vec{\delta}_u)$$

$$= \dot{\bar{\rho}} + \dot{\delta}_\rho + \bar{\rho} \nabla \cdot H\vec{r} + \bar{\rho} \nabla \cdot \vec{\delta}_u + \nabla \delta_\rho \cdot H\vec{r} + \mathcal{O}(\delta^2)$$

$$\overset{\text{Eq.(1.76)}}{=} \dot{\delta}_\rho + \bar{\rho} \nabla \cdot \vec{\delta}_u + \nabla \delta_\rho \cdot H\vec{r} + \mathcal{O}(\delta^2), \tag{1.83}$$

where we only keep terms linear in the perturbations. In the last line that the background fields solve the continuity equation (Eq. (1.76)). The Euler equation for the perturbations results in

$$0 = \left(\frac{\partial}{\partial t} + \vec{u} \cdot \nabla \right) \vec{u} = -\frac{\nabla p}{\rho_m} - \nabla \phi$$

$$= H\dot{\vec{r}} + \dot{\vec{\delta}}_u + \left(H\vec{r} + \vec{\delta}_u \right) \cdot \nabla \left(H\vec{r} + \vec{\delta}_u \right) + \frac{\nabla(\bar{p} + \delta_p)}{\bar{\rho} + \delta_\rho} + \nabla \left(\frac{1}{12M_{\text{Pl}}^2} \bar{\rho} r^2 + \delta_\phi \right)$$

$$= H\dot{\vec{r}} + \dot{\vec{\delta}}_u + H\vec{r} \cdot \nabla H\vec{r} + \vec{\delta}_u \cdot \nabla H\vec{r} + H\vec{r} \cdot \nabla \vec{\delta}_u$$

$$+ \frac{\nabla \bar{p}}{\bar{\rho}} + \frac{\nabla \delta_p}{\bar{\rho}} + \nabla \left(\frac{1}{12M_{\text{Pl}}^2} \bar{\rho} r^2 \right) + \nabla \delta_\phi + \mathcal{O}(\delta^2)$$

$$\overset{\text{Eq.(1.77)}}{=} \dot{\vec{\delta}}_u + H\vec{\delta}_u + H\vec{r} \cdot \nabla \vec{\delta}_u + \frac{\nabla \delta_p}{\bar{\rho}} + \nabla \delta_\phi + \mathcal{O}(\delta^2). \tag{1.84}$$

Finally, the Poisson equation for the fluctuations becomes

$$0 = \nabla^2 \phi - \frac{1}{2M_{\text{Pl}}^2} \rho_m$$

$$= \nabla^2 \left(\frac{1}{12M_{\text{Pl}}^2} \bar{\rho} r^2 + \delta_\phi \right) - \frac{1}{2M_{\text{Pl}}^2} (\bar{\rho} + \delta_\rho) \overset{\text{Eq.(1.78)}}{=} \nabla^2 \delta_\phi - \frac{1}{2M_{\text{Pl}}^2} \delta_\rho. \tag{1.85}$$

In analogy with Eq. (1.59) we define dimensionless fluctuations in the density field at a given place x and time t as

$$\delta(t, \vec{x}) := \frac{\rho(t, \vec{x}) - \bar{\rho}(t)}{\bar{\rho}(t)} = \frac{\delta_\rho(t, \vec{x})}{\bar{\rho}(t)}, \tag{1.86}$$

and further introduce co-moving coordinates

$$\vec{x} := \frac{a_0}{a} \vec{r} \qquad \vec{v} := \frac{a_0}{a} \vec{u} \qquad \nabla_r := \frac{a_0}{a} \nabla_x \qquad H\vec{r} \cdot \nabla_r + \frac{\partial}{\partial t} \rightarrow \frac{\partial}{\partial t}. \tag{1.87}$$

The co-moving continuity, Euler and Poisson equations then read

$$\dot{\delta} + \nabla_x \vec{\delta}_v = 0$$

$$\dot{\vec{\delta}}_v + 2H\vec{\delta}_v = -\left(\frac{a_0}{a}\right)^2 \nabla_x \left(c_s^2 \delta + \delta_\phi\right)$$

$$\nabla_x^2 \delta_\phi = \frac{1}{2M_{Pl}^2} \bar{\rho} \left(\frac{a_0}{a}\right)^2 \delta. \tag{1.88}$$

These three equations can be combined into a second order differential equation for the density fluctuations δ,

$$0 = \ddot{\delta} + \nabla_x \dot{\vec{\delta}}_v = \ddot{\delta} - \nabla_x \left[2H\vec{\delta}_v + \left(\frac{a_0}{a}\right)^2 \nabla_x \left(c_s^2 \delta + \delta_\phi\right)\right]$$

$$= \ddot{\delta} + 2H\dot{\delta} - \left(\frac{a_0}{a}\right)^2 c_s^2 \nabla_x^2 \delta - \frac{1}{2M_{Pl}^2} \bar{\rho}\delta. \tag{1.89}$$

To solve this equation, we Fourier-transform the density fluctuation and find the so-called Jeans equation

$$\delta(\vec{x}, t) = \int \frac{d^3 k}{(2\pi)^3} \hat{\delta}(\vec{k}, t) \, e^{-i\vec{k}\cdot\vec{x}}$$

$$\Rightarrow \qquad \boxed{\ddot{\hat{\delta}} + 2H\dot{\hat{\delta}} = \hat{\delta}\left[\frac{1}{2M_{Pl}^2}\bar{\rho} - \left(\frac{c_s k a_0}{a}\right)^2\right].} \tag{1.90}$$

The two competing terms in the bracket correspond to a gravitational compression of the density fluctuation and a pressure resisting this compression. The wave number for the homogeneous equation, where these terms exactly cancel defines the Jeans wave length

$$\lambda_J = \left.\frac{2\pi}{k}\right|_{homogeneous} = 2\pi \frac{a_0}{a} c_s \sqrt{\frac{2M_{Pl}^2}{\bar{\rho}}}. \tag{1.91}$$

Perturbations of this size neither grow nor get washed out by pressure. To get an idea what the Jeans length for baryons means we can compare it to the co-moving Hubble scale,

$$\frac{\lambda_J}{\frac{a_0}{a}H^{-1}} = 2\pi c_s \sqrt{\frac{2M_{Pl}^2}{\bar{\rho}}} H \overset{Eq.\,(1.81)}{=} 2\pi \sqrt{\frac{2}{3}} c_s. \tag{1.92}$$

This gives us a critical fluctuation speed of sound close to the speed of sound in relativistic matter $c_s \approx 1/\sqrt{3}$. Especially for non-relativistic matter, $c_s \ll 1$, the Jeans length is much smaller than the Hubble length and our Newtonian approach is justified.

The Jeans equation for the evolution of a non-relativistic mass or energy density can be solved in special regimes. First, for length scales much smaller than the Jeans length $\lambda \ll \lambda_J$, the Jeans equation of Eq. (1.90) becomes an equation of a damped harmonic oscillator,

$$\ddot{\hat{\delta}} + 2H\dot{\hat{\delta}} + \left(\frac{c_s k a_0}{a}\right)^2 \hat{\delta} = 0 \qquad \Rightarrow \qquad \hat{\delta}(t) \propto e^{\pm i \omega t}$$

$$\text{(non-relativistic, small structures)}, \qquad (1.93)$$

with $\omega = c_s k a_0/a$. The solutions are oscillating with decreasing amplitudes due to the Hubble friction term $2H\dot{\hat{\delta}}$. Structures with sub-Jeans lengths, $\lambda \ll \lambda_J$, therefore do not grow, but the resulting acoustic oscillations can be observed in the matter power spectrum today.

In the opposite regime, for structures larger than the Jeans length $\lambda \gg \lambda_J$, the pressure term in the Jeans equation can be neglected. The gravitational compression term can be simplified for a matter-dominated universe with $a \propto t^{2/3}$, Eq. (1.31). This gives $H = \dot{a}/a = 2/(3t)$ and it follows for the second Friedmann equation that

$$\dot{H} + H^2 = -\frac{2}{9t^2} \overset{\text{Eq. (1.80)}}{=} -\bar{\rho}\frac{1}{6M_{\text{Pl}}^2} \qquad \Rightarrow \qquad \bar{\rho} = \frac{4}{3}\frac{M_{\text{Pl}}^2}{t^2}. \qquad (1.94)$$

We can use this form to simplify the Jeans equation and solve it

$$\ddot{\hat{\delta}} + \frac{4}{3t}\dot{\hat{\delta}} - \frac{2}{3t^2}\hat{\delta} = 0 \qquad \Rightarrow \qquad \hat{\delta} = At^{2/3} + \frac{B}{t}$$

$$\propto t^{2/3} \qquad \text{growing mode}$$

$$=: \frac{a}{a_0}\hat{\delta}_0 \qquad \text{using } a \propto t^{2/3}$$

$$\text{(non-relativistic, large structures)}. \qquad (1.95)$$

We can use this formula for the growth as a function of the scale factor to link the density perturbations at the time of photon decoupling to today. For this we quote that at photon decoupling we expect $\hat{\delta}_{\text{dec}} \approx 10^{-5}$, which gives us for today

$$\boxed{\hat{\delta}_0 = \frac{\hat{\delta}_{\text{dec}}}{a_{\text{dec}}} \overset{\text{Eq. (1.57)}}{=} 1100\,\hat{\delta}_{\text{dec}} \approx 1\%}. \qquad (1.96)$$

We can compare this value with the results from numerical N-body simulations and find that those simulations prefer much larger values $\hat{\delta}_0 \approx 1$. In other words, the smoothness of the CMB shows that perturbations in the photon-baryon fluid alone cannot account for the cosmic structures observed today. One way to improve the situation is to introduce a dominant non-relativistic matter component with a negligible pressure term, defining the main properties of cold dark matter.

Until now our solutions of the Jeans equation rely on the assumption of non-relativistic matter domination. For relativistic matter with $a \propto t^{1/2}$ the growth of density perturbations follows a different scaling. Following Eq. (1.32) we use $H = t/2$ and assume $H^2 \gg 4\pi G \bar{\rho}$, such that the Jeans equation becomes

$$\ddot{\hat{\delta}} + \frac{\dot{\hat{\delta}}}{t} = 0 \quad \Rightarrow \quad \hat{\delta} = A + B \log t \qquad \text{(relativistic, small structures)}.$$

$$(1.97)$$

This growth of density perturbations is much weaker than for non-relativistic matter.

Finally, we have to consider relativistic density perturbations larger than the Hubble scale, $\lambda \gg a/a_0 H$. In this case a Newtonian treatment is no longer justified and we only quote the result of the full calculation from general relativity, which gives a scaling

$$\hat{\delta} = \left(\frac{a}{a_0}\right)^2 \hat{\delta}_0 \qquad \text{(relativistic, large structures)}. \qquad (1.98)$$

Together with Eqs. (1.93), (1.95) and (1.97) this gives us the growth of structures as a function of the scale parameter for non-relativistic and relativistic matter and for small as well as large structures. Radiation pressure in the photon-baryon fluid prevents the growth of small baryonic structures, but baryon-acoustic oscillations on smaller scales predicted by Eq. (1.93) can be observed. Large structures in a relativistic, radiation-dominated universe indeed expand rapidly. Later in the evolution of the Universe, non-relativistic structures come close to explaining the matter density in the CMB evolving to today as we see it in numerical simulations, but require a dominant additional matter component.

Similar to the variations of the cosmic microwave photon temperature we can expand our analysis of the matter density from the central value to its distribution with different sizes or wave numbers. To this end we define the matter power spectrum $P(k)$ in momentum space as

$$\langle \hat{\delta}(\vec{k})^* \hat{\delta}(\vec{k}') \rangle = (2\pi)^3 \delta(\vec{k} - \vec{k}') P(k). \qquad (1.99)$$

As before, we can link k to a wave length $\lambda = 2\pi/k$. For the scaling of the initial power spectrum the proposed relation by Harrison and Zel'dovich is

$$P(k) \propto k^n = \left(\frac{2\pi}{\lambda}\right)^n. \qquad (1.100)$$

From observations we know that $n \gg 1$ leads to an increase in small-scale structures and as a consequence to too many black holes. We also know that for $n \ll 1$, large structures like super-clusters dominate over smaller structures like galaxies, again contradicting observations. Based on this, the exponent was originally predicted to be $n = 1$, in agreement with standard inflationary cosmology. However, the global CMB analysis by PLANCK quoted in Eq. (1.75) gives

$$n = 0.9645 \pm 0.0049. \tag{1.101}$$

We can solve this slight disagreement by considering perturbations of different size separately. First there are small perturbations (large k), which enter the horizon of our expanding Universe during the radiation-dominated era and hardly grow until matter-radiation equality. Second, there are large perturbations with (small k), which only enter the horizon during matter domination and never stop growing. This freezing of the growth before matter domination is called the Meszaros effect. Following Eq. (1.98) the relative suppression in their respective growth between the entering into the horizon and the radiation-matter equality is given by a the correction factor relative to Eq. (1.100) with $n = 1$,

$$P(k) \propto k \left(\frac{a_{\text{enter}}}{a_{\text{eq}}} \right)^2. \tag{1.102}$$

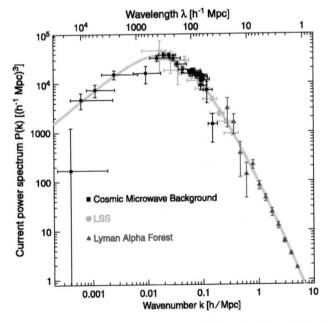

Fig. 1.4 Best fit of today's matter power spectrum ($a_0 = 1$) from Max Tegmark's lecture notes [3]

We are interested in the wavelength of a mode that enters right at matter-radiation equality and hence is the first mode that never stops growing. Assuming the usual scaling $\Omega_m/\Omega_r \propto a$ and we first find

$$\frac{a_{eq}}{a_0} = \frac{\dfrac{\Omega_m(a_{eq})}{\Omega_r(a_{eq})}}{\dfrac{\Omega_m(a_0)}{\Omega_r(a_0)}} = \frac{\Omega_r(a_0)}{\Omega_m(a_0)} \approx 3 \cdot 10^{-4}, \tag{1.103}$$

again from PLANCK measurements. This allows us to integrate the co-moving distance of Eq. (1.36). The lower and upper limit of integration is $a = 0$ and $a = a_{eq} = 3 \cdot 10^{-4}$, respectively. For these values of $a \ll 1$ the relativistic matter dominates the universe, as can be seen in Fig. 1.1. In this range the integrand of Eq. (1.36) is approximately

$$\left[\Omega_m(t_0)a(t) + \Omega_r(t_0) + \Omega_\Lambda a(t)^4 - (\Omega_t(t_0) - 1)a(t)^2\right]^{-1/2} \approx \frac{1}{\sqrt{\Omega_r(t_0)}}. \tag{1.104}$$

This is true even for $\Omega_\Lambda(t_0) > \Omega_m(t_0) > \Omega_r(t_0)$ today. We can use Eq. (1.104) and write

$$d_{eq}^c \approx \int_0^{a_{eq}} da \, \frac{1}{H_0\sqrt{\Omega_r(t_0)}} = \frac{a_{eq}}{H_0\sqrt{\Omega_r(t_0)}}$$

$$\overset{\text{Eq. (1.103)}}{=} \frac{3 \cdot 10^{-4}}{70\frac{\text{km}}{\text{s Mpc}}\sqrt{0.28 \times 3 \cdot 10^{-4}}} = 4.7 \cdot 10^{-4}\frac{\text{Mpc s}}{\text{km}}$$

$$\Rightarrow \quad \lambda_{eq} = c\,d_{eq}^c \approx 3 \cdot 10^5 \frac{\text{km}}{\text{s}} \times 4.7 \cdot 10^{-4}\frac{\text{Mpc s}}{\text{km}} = 140\,\text{Mpc}. \tag{1.105}$$

This means that the growth of structures with a size of at least 140 Mpc never stalls, while for smaller structures the Meszaros effect leads to a suppressed growth. The scaling of λ_{eq} in the radiation dominated era in dependence of the scale factor is given by $\lambda_{eq} \propto a_{eq}c/H_0$. The co-moving wavenumber is defined as $k = 2\pi/\lambda$ and therefore $k_{eq} \approx 0.05/\text{Mpc}$. Using this scaling, $a \propto 1/k$, the power spectrum scales as

$$P(k) \propto k \left(\frac{a_{enter}}{a_{eq}}\right)^2 = \begin{cases} k & k < k_{eq} \text{ or } \lambda > 120\,\text{Mpc} \\ \dfrac{1}{k^3} & k > k_{eq} \text{ or } \lambda < 120\,\text{Mpc}. \end{cases} \tag{1.106}$$

The measurement of the power spectrum shown in Fig. 1.4 confirms these two regimes.

Even if pressure can be neglected for cold, collision-less dark matter, its perturbations cannot collapse towards arbitrary small scales because of the non-zero velocity dispersion. Once the velocity of dark matter particles exceeds the escape velocity of a density perturbation, they will stream away before they can be gravitationally bound. This phenomenon is called free streaming and allows us to derive more properties of the dark matter particles from the matter power spectrum. To this end we generalize the Jeans equation of Eq. (1.90) to

$$\ddot{\hat{\delta}} + 2H\dot{\hat{\delta}} = \hat{\delta}\left[\frac{1}{2M_{\mathrm{Pl}}^2}\bar{\rho} - \left(c_s^{\mathrm{eff}}\frac{ka_0}{a}\right)^2\right],\tag{1.107}$$

where in the term counteracting the gravitational attraction the speed of sound is replaced by an effective speed of sound c_s^{eff}, whose precise form depends on the properties of the dark matter. We show predictions for different dark matter particles in Fig. 1.5 [4]:

– for cold dark matter with

$$(c_s^{\mathrm{eff}})^2 = \frac{1}{m^2}\frac{\int dp\, p^2 f(p)}{\int dp\, f(p)}\tag{1.108}$$

the speed of sound is replaced by the non-relativistic velocity distribution. This results in $c_s^{\mathrm{eff}} \ll c_s$ and the cold dark matter Jeans length allows for halo structures as small as stars or planets. The dominant dark matter component in Fig. 1.4 is cold collision-less dark matter and all lines in Fig. 1.5 is normalized to this the power spectrum;
– for warm dark matter with

$$c_s^{\mathrm{eff}} = \frac{T}{m}\tag{1.109}$$

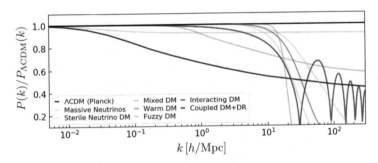

Fig. 1.5 Sketch of the matter power spectrum for different dark matter scenarios normalized to the ΛCDM power spectrum. Figure from Ref. [4]

the effective speed of sound is a function of temperature and mass. Warm dark matter is faster than cold dark matter and the effective speed of sound is larger. As a result, small structures are washed out as indicated by the blue line, because the free streaming length for warm dark matter is larger than for cold dark matter;

– sterile neutrinos which we will introduce in Sect. 2.1 feature

$$(c_s^{\text{eff}})^2 = \frac{1}{m^2} \frac{\int dp \, p^2 f(p)}{\int dp \, f(p)}. \tag{1.110}$$

They are a special case of warm dark matter, but the result of the integral depends on the velocity distribution, which is model-dependent. In general a suppression of small scale structures is expected and the resulting normalized power spectrum should end up between the two cyan lines;

– light, non-relativistic dark matter or fuzzy dark matter which we will discuss in Sect. 2.2 gives

$$c_s^{\text{eff}} = \frac{k}{m}. \tag{1.111}$$

The effective speed of sound depends on k, leading to an even stronger suppression of small scale structures. The normalized power spectrum is shown in turquoise;

– for mixed warm and cold dark matter with

$$(c_s^{\text{eff}})^2 = \frac{T^2}{m^2} - \frac{\bar{\rho}}{2M_{\text{Pl}}^2} \frac{a}{a_0 k} \frac{\hat{\delta}_C}{\hat{\delta}} \tag{1.112}$$

the power spectrum is suppressed. Besides a temperature-dependent speed of sound for the warm dark matter component, a separate gravitational term for the cold dark matter needs to be added in the Jeans equation. Massive neutrinos are a special case of this scenario and in turn the power spectrum can be used to constrain SM neutrino masses;

– finally, self-interacting dark matter with the distinctive new term

$$(c_s^{\text{eff}})^2 = c_s^{\text{dark}} + R(\dot{\hat{\delta}} - \dot{\hat{\delta}}_\chi) \frac{a}{a_0 k} \frac{1}{\hat{\delta}} \tag{1.113}$$

covers models from a dark force (dark radiation) to multi-component dark matter that could form dark atoms. Besides a potential dark sound speed, the Jeans equation needs to be modified by an interaction term. The effects on the power spectrum range from dark acoustic oscillations to a suppression of structures at multiple scales.

References

1. Baumann, D.: Cosmology. http://cosmology.amsterdam/education/cosmology/
2. Planck Collaboration: Planck 2015 results. XI. CMB power spectra, likelihoods, and robustness of parameters. Astron. Astrophys. (submitted). arXiv:1507.02704
3. Tegmark, M.: Measuring Space-Time: From Big Bang to Black Holes. Lect. Notes Phys. 646, 169 (2004). arXiv:astro-ph/0207199
4. Schneider, A.: Cosmic Structure Formation and Dark Matter. Talk at workshop "Dark Matter at the Dawn of a Discovery?", Heidelberg, 2018. https://indico.cern.ch/event/678386/contributions/2893445/attachments/1631142/2600258/Schneider_Heidelberg2018.pdf

Chapter 2
Relics

After we understand the relic photons in the Universe, we can focus on a set of different other relics, including the first dark matter candidates. For those the main question is to explain the observed value of $\Omega_\chi h^2 \approx 0.12$. Before we will eventually turn to thermal production of massive dark matter particles, we can use a similar approach as for the photons for relic neutrinos. Furthermore, we will look at ways to produce dark matter during the thermal history of the Universe, without relying on the thermal bath.

2.1 Relic Neutrinos

In analogy to photon decoupling, just replacing the photon–electron scattering rate given in Eq. (1.49) by the much larger neutrino–electron scattering rate, we can also compute today's neutrino background in the Universe. At the point of decoupling the neutrinos decouple from the electrons and photons, but they will also lose the ability to annihilate among themselves through the weak interaction. A well-defined density of neutrinos will therefore freeze out of thermal equilibrium. The two processes

$$\bar{\nu}_e e^- \to \bar{\nu}_e e^- \qquad \text{and} \qquad \nu_e \bar{\nu}_e \to e^+ e^- \tag{2.1}$$

are related through a similar scattering rate $\sigma_{\nu e}$ given in Eq. (1.49).

Because the neutrino scattering cross section is small, we expect the neutrinos to decouple earlier. It turns out that this happens before nucleosynthesis. This means that for the relic neutrinos the electrons are the relevant degrees of freedom to compute the decoupling temperature. With the cross section given in Eq. (1.49) the

© Springer Nature Switzerland AG 2019
M. Bauer, T. Plehn, *Yet Another Introduction to Dark Matter*,
Lecture Notes in Physics 959, https://doi.org/10.1007/978-3-030-16234-4_2

interaction rates for relativistic neutrino–electron scattering is

$$\Gamma_\nu = \sigma_{\nu e} \nu n_\nu = \sigma_{\nu e} \nu n_e \overset{\text{Eq. (1.40)}}{\approx} \frac{\pi \alpha^2 T^2}{s_w^4 m_W^4} \frac{3\zeta_3}{4\pi^2} gT^3 = \frac{3\zeta_3}{4\pi s_w^4} g\alpha^2 \frac{T^5}{m_W^4} . \tag{2.2}$$

With only one generation of neutrinos in the initial state and a purely left-handed coupling the number of relativistic degrees of freedom relevant for this scattering process is $g = 1$.

Just as for the photons, we first compute the decoupling temperature. To link the interaction rate to the Hubble constant, as given by Eq. (1.47), we need the effective number of degrees of freedom in the thermal bath. It now includes electrons, positrons, three generations of neutrinos, and photons

$$g_{\text{eff}}(T_{\text{dec}}) = \frac{7}{8} (2 + 2 + 3 \times 2) + 2 = 10.75 . \tag{2.3}$$

With Eq. (1.47) and in analogy to Eq. (1.55) we find

$$\frac{\Gamma_\nu}{H} = \frac{3\zeta_3 g\alpha}{4\pi^2 s_w^4} \frac{T^5}{m_W^4} \frac{\sqrt{90}M_{\text{Pl}}}{\pi} \frac{1}{\sqrt{g_{\text{eff}}(T)} T^2}$$

$$= \frac{9\sqrt{10}\,\zeta_3}{4\pi^2 s_w^4} \alpha^2 \frac{g}{\sqrt{g_{\text{eff}}(T)}} \frac{M_{\text{Pl}} T^3}{m_W^4} \overset{!}{=} 1$$

$$\Leftrightarrow T_{\text{dec}} = \left(\frac{4\pi^2 s_w^4}{9\sqrt{10}\,\zeta_3} \frac{m_W^4}{M_{\text{Pl}}} \frac{\sqrt{g_{\text{eff}}(T_{\text{dec}})}}{\alpha^2 g} \right)^{1/3} \approx (1 \dots 10) \text{ MeV} . \tag{2.4}$$

The relativistic neutrinos decouple at a temperature of a few MeV, before nucleosynthesis. From the full Boltzmann equation we would get $T_{\text{dec}} \approx 1$ MeV, consistent with our approximate computation.

Now that we know how the neutrinos and photons decouple from the thermal bath, we follow the electron-neutrino-photon system from the decoupling phase to today, dealing with one more relevant effect. First, right after the neutrinos decouple around $T_{\text{dec}} \approx 1$ MeV, the electron with a mass of $m_e = 0.5$ MeV will drop from the relevant relativistic degrees of freedom; in the following phase the electrons will only serve as a background for the photon. For the evolution to today we only have

$$g_{\text{eff}}(T_{\text{dec}} \dots T_0) = \frac{7}{8} \times 6 + 2 = 7.25 \tag{2.5}$$

relativistic degrees of freedom. The decoupling of the massive electron adds one complication: in the full thermodynamic calculation we need to assume that their entropy is transferred to the photons, the only other particles still in equilibrium. We only quote the corresponding result from the complete calculation: because the entropy in the system should not change in this electron decoupling process, the temperature of the photons jumps from

$$T_\gamma = T_\nu \qquad \rightarrow \qquad T_\gamma = \left(\frac{11}{4}\right)^{1/3} T_\nu \,. \tag{2.6}$$

If the neutrino and photon do not have the same temperature we can use Eqs. (1.43) and (1.44) to obtain the combined relativistic matter density at the time of neutrino decoupling,

$$\rho_r(T) = \frac{\pi^2}{30} g_{\text{eff}}(T) \, T^4 = \frac{\pi^2}{30} \left(2 \frac{T_\gamma^4}{T^4} + \frac{7}{8} \, 6 \, \frac{T_\nu^4}{T^4}\right) T^4$$

$$\Rightarrow \rho_r(T_\gamma) = \frac{\pi^2}{30} \left(2 + \frac{21}{4} \left(\frac{4}{11}\right)^{4/3}\right) T_\gamma^4 = \frac{\pi^2}{30} \, 3.4 \, T_\gamma^4 \,, \tag{2.7}$$

or $g_{\text{eff}}(T) = 3.4$. This assumes that we measure the current temperature of the Universe through the photons. Assuming a constant suppression of the neutrino background, its temperature and the total relativistic energy density today are

$$T_{0,\nu} = 1.7 \cdot 10^{-4} \, \text{eV} \qquad \text{and} \qquad \rho_r(T_0) = \frac{\pi^2}{30} \, 3.4 \, T_{0,\gamma}^4 = 1.1 \, T_{0,\gamma}^4 \,. \tag{2.8}$$

From the composition in Eq. (2.7) we see that the current relativistic matter density of the Universe is split roughly $60 - 40$ between the photons at $T_{0,\gamma} = 2.4 \cdot 10^{-4} \, \text{eV}$ and the neutrinos at $T_{0,\nu} = 1.7 \cdot 10^{-4} \, \text{eV}$. The normalized relativistic relic density today becomes

$$\boxed{\Omega_r(t_0)h^2 = \frac{\rho_r(T_0)h^2}{3M_{\text{Pl}}^2 H_0^2} = 0.54 \left(\frac{2.4 \cdot 10^{-4} \, \text{eV}}{2.5 \cdot 10^{-3} \, \text{eV}}\right)^4 = 4.6 \cdot 10^{-5}} \,. \tag{2.9}$$

Note that for this result we assume that the neutrino mass never plays a role in our calculation, which is not at all a good approximation.

We are now in a position to answer the question whether a massive, stable fourth neutrino could explain the observed dark matter relic density. With a moderate mass, this fourth neutrino decouples in a relativistic state. In that case we can relate its number density to the photon temperature through Eq. (2.7),

$$n_\nu(T) \stackrel{\text{Eq. (1.40)}}{=} \frac{3}{4} \frac{2\zeta_3}{\pi^2} T_\nu^3 \stackrel{\text{Eq. (2.6)}}{=} \frac{6\zeta_3}{11\pi^2} T_\gamma^3 \,. \tag{2.10}$$

With decreasing temperature a heavy neutrino will at some point become non-relativistic. This means we use the non-relativistic relation to compute its energy density today,

$$\rho_\nu(T_0) = m_\nu n_\nu(T_0) = m_\nu \frac{6\,\zeta_3}{11\pi^2} T_{0,\gamma}^3$$

$$\Rightarrow \quad \Omega_\nu h^2 = m_\nu \frac{6\,\zeta_3}{11\pi^2} T_{0,\gamma}^3 \frac{h^2}{3M_{\rm Pl}^2 H_0^2} = \frac{m_\nu}{30} \frac{(2.4 \cdot 10^{-4})^3}{(2.5 \cdot 10^{-3})^4} \frac{1}{\rm eV} = \frac{m_\nu}{85\,\rm eV} \, .$$

(2.11)

For an additional, heavy neutrino to account for the observed dark matter we need to require

$$\Omega_\nu h^2 \overset{!}{=} \Omega_\chi h^2 \approx 0.12 \qquad \Leftrightarrow \qquad m_\nu \approx 10\,{\rm eV}\,.$$

(2.12)

This number for hot neutrino dark matter is not unreasonable, as long as we only consider the dark matter relic density today. The problem appears when we study the formation of galaxies, where it turns out that dark matter relativistic at the point of decoupling will move too fast to stabilize the accumulation of matter. We can look at Eq. (2.12) another way: if all neutrinos in the Universe add to more than this mass value, they predict hot dark matter with a relic density more than then entire dark matter in the Universe. This gives a stringent upper bound on the neutrino mass scale.

2.2 Cold Light Dark Matter

Before we introduce cold and much heavier dark matter, there is another scenario we need to discuss. Following Eq. (2.12) a new neutrino with mass around $10\,{\rm eV}$ could explain the observed relic density. The problem with thermal neutrino dark matter is that it would be relativistic at the wrong moment of the thermal history, causing serious issues with structure formation as discussed in Sect. 1.5. The obvious question is if we can modify this scenario such that light dark matter remains non-relativistic. To produce such light cold dark matter we need a non-thermal production process.

We consider a toy model for light cold dark matter with a spatially homogeneous but time-dependent complex scalar field $\phi(t)$ with a potential V. For the latter, the Taylor expansion is dominated by a quadratic mass term m_ϕ. Based on the invariant action with the additional determinant of the metric g, describing the expanding Universe, the Lagrangian for a single complex scalar field reads

$$\frac{1}{\sqrt{|g|}} \mathscr{L} = (\partial^\mu \phi^*)(\partial_\mu \phi) - V(\phi) = (\partial^\mu \phi^*)(\partial_\mu \phi) - m_\phi^2 \, \phi^* \phi \, .$$

(2.13)

Just as a side remark, the difference between the Lagrangians for real and complex scalar fields is a set of factors $1/2$ in front of each term. In our case the equation of motion for a spatially homogeneous field $\phi(t)$ is

$$
\begin{aligned}
0 &= \partial_t \left(\frac{\partial \mathcal{L}}{\partial (\partial_t \phi^*)} \right) - \frac{\partial \mathcal{L}}{\partial \phi^*} \\
&= \partial_t \left(\sqrt{|g|}\, \partial_t \phi \right) + \sqrt{|g|}\, m_\phi^2 \phi \\
&= (\partial_t \sqrt{|g|})\, (\partial_t \phi) + \sqrt{|g|}\, \partial_t^2 \phi + \sqrt{|g|}\, m_\phi^2 \phi \\
&= \sqrt{|g|} \left[\frac{(\partial_t \sqrt{|g|})}{\sqrt{|g|}} (\partial_t \phi) + \partial_t^2 \phi + m_\phi^2 \phi \right] .
\end{aligned} \tag{2.14}
$$

For example from Eq. (1.13) we know that in flat space ($k = 0$) the determinant of the metric is $|g| = a^6$, giving us

$$
0 = \frac{(\partial_t a^3)}{a^3} (\partial_t \phi) + \partial_t^2 \phi + m_\phi^2 \phi = \frac{3\dot{a}}{a}\, \dot{\phi} + \ddot{\phi} + m_\phi^2 \phi . \tag{2.15}
$$

Using the definition of the Hubble constant in Eq. (1.14) we find that the expansion of the Universe is responsible for the friction term in

$$
\boxed{\ddot{\phi}(t) + 3H\dot{\phi}(t) + m_\phi^2 \phi(t) = 0} . \tag{2.16}
$$

We can solve this equation for the evolving Universe, described by a decreasing Hubble constant with increasing time or decreasing temperature, Eq. (1.47). If for each regime we assume a constant value of H—an approximation we need to check later—and find

$$
\phi(t) = e^{i\omega t} \Rightarrow \dot{\phi}(t) = i\omega \phi(t) \Rightarrow \ddot{\phi}(t) = -\omega^2 \phi(t)
$$

$$
\Rightarrow -\omega^2 + 3i H\omega + m_\phi^2 = 0
$$

$$
\Rightarrow \omega = \frac{3i}{2} H \pm \sqrt{-\frac{9}{4}H^2 + m_\phi^2} . \tag{2.17}
$$

This functional form defines three distinct regimes in the evolution of the Universe:

- In the early Universe $H \gg m_\phi$ the two solutions are $\omega = 0$ and $\omega = 3iH$. The scalar field value is a combination of a constant mode and an exponentially decaying mode.

$$
\phi(t) = \phi_1 + \phi_2\, e^{-3Ht} \xrightarrow{\text{time evolution}} \phi_1 . \tag{2.18}
$$

The scalar field very rapidly settles in a constant field value and stays there. There is no good reason to assume that this constant value corresponds to a minimum of the potential. Due to the Hubble friction term in Eq. (2.16), there is simply no time for the field to evolve towards another, minimal value. This behavior gives the process its name, misalignment mechanism. For our dark matter considerations we are interested in the energy density. Following the virial theorem we assume that the total energy density stored in our spatially constant field is twice the average potential energy $V = m_\phi^2 |\phi|^2/2$. After the rapid decay of the exponential contribution this means

$$\rho(t) \to m_\phi^2 \, \phi_1^2 \, . \tag{2.19}$$

– A transition point in the evolution of the universe occurs when the evolution of the field ϕ switches from the exponential decay towards a constant value ϕ_1 to an oscillation mode. If we identify the oscillation modes of the field ϕ with a dark matter degree of freedom, this point in the thermal history defines the production of cold, light dark matter,

$$H_{\text{prod}} \approx m_\phi \qquad \Leftrightarrow \qquad \omega \approx \frac{3i}{2} H_{\text{prod}} \, . \tag{2.20}$$

– For the late Universe $H \ll m_\phi$ we expand the complex eigen-frequency one step further,

$$\omega = \frac{3i}{2} H \pm m_\phi \sqrt{1 - \frac{9H^2}{4m_\phi^2}} \approx \frac{3i}{2} H \pm m_\phi \left(1 - \frac{9H^2}{8m_\phi^2}\right) \approx \pm m_\phi + \frac{3i}{2} H \, . \tag{2.21}$$

The leading time dependence of the scalar field is an oscillation. The subleading term, suppressed by H/m_ϕ, describes an exponentially decreasing field amplitude,

$$\phi(t) = \phi_3 \, e^{\pm i m_\phi t} \, e^{-3H/2t} \, . \tag{2.22}$$

A modification of the assumed constant H value changes the rapid decay of the amplitude, but should not affect these main features. We can understand the physics of this late behavior when we compare it to the variation of the scale factor for constant H given in Eq. (1.14),

$$H = \frac{\dot{a}(t)}{a(t)} \quad \Rightarrow \quad a(t) \propto e^{Ht}$$

$$\Rightarrow \quad \rho(t) \stackrel{\text{Eq. (2.22)}}{=} m_\phi^2 \, |\phi_3|^2 \, e^{-3Ht} \propto \frac{1}{a(t)^3}$$

$$\Leftrightarrow \quad \boxed{\frac{\rho(t)}{\rho_0} \propto \frac{a_0^3}{a(t)^3}} \, . \tag{2.23}$$

The energy density of the scalar field in this late regime is inversely proportional to the space volume element in the expanding Universe. This relation is exactly what we expect from a non-relativistic relic without any interaction or quantum effects.

Next, we can use Eq. (2.23) combined with the assumption of constant H to approximately relate the dark matter relic densities at the time of production with today $a_0 = 1$,

$$0.12 \approx \Omega_\chi h^2 = \frac{\rho_\chi}{\rho_c} h^2 \stackrel{\text{Eq. (2.19)}}{=} \frac{m_\phi^2 \phi^2(t_0)}{(2.5 \cdot 10^{-3} \, \text{eV})^4} h^2$$

$$\stackrel{\text{Eq. (2.23)}}{=} \frac{m_\phi^2 \phi^2(t_{\text{prod}})}{(2.5 \cdot 10^{-3} \, \text{eV})^4} \frac{a(t_{\text{prod}})^3}{a_0^3} h^2 . \tag{2.24}$$

Using our thermodynamic result $a(T) \propto 1/T$ from Eq. (1.33) and the approximate relation between the Hubble parameter and the temperature at the time of production we find

$$\sqrt{0.12} = \frac{m_\phi \phi(t_{\text{prod}})}{(2.5 \cdot 10^{-3} \, \text{eV})^2} \frac{T_0^{3/2}}{T_{\text{prod}}^{3/2}} h \stackrel{\text{Eq. (1.47)}}{\approx} \frac{m_\phi \phi(t_{\text{prod}})}{(2.5 \cdot 10^{-3} \, \text{eV})^2} \frac{T_0^{3/2}}{(H_{\text{prod}} M_{\text{Pl}})^{3/4}} h .$$

$$\tag{2.25}$$

Moreover, from Eq. (2.20) we know that the Hubble constant at the time of dark matter production is $H_{\text{prod}} \sim m_\phi$. This leads us to the relic density condition for dark matter produced by the misalignment mechanism,

$$0.35 = \frac{m_\phi \phi(t_{\text{prod}})}{(2.5 \cdot 10^{-3} \, \text{eV})^2} \frac{T_0^{3/2}}{(m_\phi M_{\text{Pl}})^{3/4}} h \tag{2.26}$$

$$\Leftrightarrow \quad m_\phi \phi(t_{\text{prod}}) = \frac{1}{2} \frac{(2.5 \cdot 10^{-3} \, \text{eV})^2}{(2.4 \cdot 10^{-4} \, \text{eV})^{3/2}} (m_\phi M_{\text{Pl}})^{3/4} \approx (m_\phi M_{\text{Pl}})^{3/4} \, \text{eV}^{1/2} .$$

This is the general relation between the mass of a cold dark matter particle and its field value, based on the observed relic density. If the misalignment mechanism should be responsible for today's dark matter, inflation occurring after the field ϕ has picked its non-trivial starting value will have to give us the required spatial homogeneity. This is exactly the same argument we used for the relic photons in Sect. 1.4. We can then link today's density to the density at an early starting point through the evolution sketched above.

Before we illustrate this behavior with a specific model we can briefly check when and why this dark matter candidate is non-relativistic. If through some unspecified quantization we identify the field oscillations of ϕ with dark matter particles, their non-relativistic velocity is linked to the field value ϕ through the quantum mechanical definition of the momentum operator,

$$v = \frac{\hat{p}}{m} \propto \frac{\partial \phi}{\partial x} \,, \tag{2.27}$$

assuming an appropriate normalization by the field value ϕ. It can be small, provided we find a mechanism to keep the field ϕ spatially constant. What is nice about this model for cold, light dark matter is that it requires absolutely no particle physics calculations, no relativistic field theory, and can always be tuned to work.

2.3 Axions

The best way to guarantee that a particle is massless or light is through a symmetry in the Lagrangian of the quantum field theory. For example, if the Lagrangian for a real spin-0 field $\phi(x) \equiv a(x)$ is invariant under a constant shift $a(x) \rightarrow a(x) + c$, a mass term $m_a^2 a^2$ breaks this symmetry. Such particles, called Nambu-Goldstone bosons, appear in theories with broken global symmetries. Because most global symmetry groups are compact or have hermitian generators and unitary representations, the Nambu-Goldstone bosons are usually CP-odd.

We illustrate their structure using a complex scalar field transforming under a $U(1)$ rotation, $\phi \rightarrow \phi e^{ia/f_a}$. A vacuum expectation value $\langle \phi \rangle = f_a$ leads to spontaneous breaking of the $U(1)$ symmetry, and the Nambu-Goldstone boson a will be identified with the broken generator of the phase. If the complex scalar has couplings to chiral fermions ψ_L and ψ_R charged under this $U(1)$ group, the Lagrangian includes the terms

$$\mathcal{L} \supset i\overline{\psi}_L \gamma^\mu \partial_\mu \psi_L + i\overline{\psi}_R \gamma^\mu \partial_\mu \psi_R - y\,\phi\,\overline{\psi}_R \psi_L + \text{h.c.} \tag{2.28}$$

We can rewrite the Yukawa coupling such that after the rotation the phase is absorbed in the definition of the fermion fields,

$$y\,\phi\,\overline{\psi}_R \psi_L \rightarrow y\,f_a\,\overline{\psi}_R e^{ia/f_a}\,\psi_L \equiv y\,f_a\,\overline{\psi}'_R \psi'_L$$

$$\text{with} \quad \psi'_{R,L} = e^{\mp ia/(2f_a)} \psi_{R,L} \,. \tag{2.29}$$

This gives us a fermion mass $m_\psi = y f_a$. In the new basis the kinetic terms read

$$i\overline{\psi}_L \gamma^\mu \partial_\mu \psi_L + i\overline{\psi}_R \gamma^\mu \partial_\mu \psi_R$$

$$= i\,\overline{\psi}'_L e^{-ia/(2f_a)} \gamma^\mu \partial_\mu e^{ia/(2f_a)} \psi'_L$$

$$+ i\,\overline{\psi}'_R e^{ia/(2f_a)} \gamma^\mu \partial_\mu e^{-ia/(2f_a)} \psi'_R$$

$$= i\,\overline{\psi}'_L \gamma^\mu \left(\partial_\mu + i\frac{(\partial_\mu a)}{2f_a} \right) \psi'_L$$

$$+ i\,\overline{\psi}'_R \gamma^\mu \left(\partial_\mu - i\frac{(\partial_\mu a)}{2f_a} \right) \psi'_R + \mathcal{O}(f_a^{-2})$$

$$= i\,\overline{\psi} \gamma^\mu \partial_\mu \psi + \frac{(\partial_\mu a)}{2f_a} \overline{\psi} \gamma^\mu \gamma_5\, \psi + \mathcal{O}(f_a^{-2})\,, \tag{2.30}$$

where in the last line we define the four-component spinor $\psi \equiv (\psi'_L, \psi'_R)$. The derivative coupling and the axial structure of the new particle a are evident. Other structures arise if the underlying symmetry is not unitary, as is the case for space-time symmetries for which the group elements can be written as $e^{\alpha/f}$ and a calculation analogous to Eq. (2.30) leads to scalar couplings. The Nambu-Goldstone boson of the scale symmetry, the dilaton, is an example of such a case.

Following Eq. (2.30), the general shift-symmetric Lagrangian for such a CP-odd pseudo-scalar reads

$$\mathcal{L} = \frac{1}{2}(\partial_\mu a)\,(\partial^\mu a) + \frac{a}{f_a}\frac{\alpha_s}{8\pi} G^a_{\mu\nu}\widetilde{G}^{a\,\mu\nu} + c_\gamma \frac{a}{f_a}\frac{\alpha}{8\pi} F_{\mu\nu}\widetilde{F}^{\mu\nu}$$

$$+ \frac{(\partial_\mu a)}{2f_a} \sum_\psi c_\psi\, \overline{\psi}\gamma^\mu \gamma_5 \psi\,. \tag{2.31}$$

The coupling to the Standard Model is mediated by a derivative interactions to the (axial) current of all SM fermions. Here $\widetilde{F}_{\mu\nu} = \epsilon_{\mu\nu\rho\tau} F^{\rho\tau}/2$ and correspondingly $\widetilde{G}_{\mu\nu}$ are the dual field-strength tensors. This setup has come to fame as a possible solution to the so-called strong CP-problem. In QCD, the dimension-4 operator

$$\frac{\theta_{\rm QCD}}{8\pi} G^a_{\mu\nu}\widetilde{G}^{a\,\mu\nu} \tag{2.32}$$

respects the $SU(3)$ gauge symmetry, but would induce observable CP-violation, for example a dipole moment for the neutron. Note that it is non-trivial that this operator cannot be ignored, because it looks like a total derivative, but due to the topological structure of $SU(3)$, it doesn't vanish. The non-observation of a neutron dipole moment sets very strong constraints on $\theta_{\rm QCD} < 10^{-10}$. This almost looks like this operator shouldn't be there and yet there is no symmetry in the Standard Model that forbids it.

Combining the gluonic operators in Eqs. (2.31) and (2.32) allows us to solve this problem

$$\mathcal{L} = \frac{1}{2}(\partial_\mu a)\,(\partial^\mu a) + \frac{\alpha_s}{8\pi} \left(\frac{a}{f_a} - \theta_{\rm QCD} \right) G^a_{\mu\nu}\widetilde{G}^{a\,\mu\nu} + c_\gamma \frac{a}{f_a}\frac{\alpha}{8\pi} F_{\mu\nu}\widetilde{F}^{\mu\nu}$$

$$+ \frac{(\partial_\mu a)}{2f_a} \sum_\psi c_\psi\, \overline{\psi}\gamma^\mu \gamma_5 \psi\,. \tag{2.33}$$

With this ansatz we can combine the θ-parameter and the scalar field, such that after quarks and gluons have formed hadrons, we can rewrite the corresponding effective Lagrangian including the terms

$$\mathscr{L}_{\text{eff}} \supset \frac{1}{2}(\partial_\mu a)\,(\partial^\mu a) - \frac{1}{2}\kappa^2 \left(\theta_{\text{QCD}} - \frac{a}{f_a}\right)^2 - \lambda_a \left(\theta_{\text{QCD}} - \frac{a}{f_a}\right)^4 + \mathcal{O}(f_a^{-6})\,.$$

(2.34)

The parameters κ and λ_a depend on the QCD dynamics. This contribution provides a potential for a with a minimum at $\langle a \rangle / f_a = \theta_{\text{QCD}}$. In other words, the shift symmetry has eliminated the CP-violating gluonic term from the theory. Because of its axial couplings to matter fields, the field a is called axion.

The axion would be a bad dark matter candidate if it was truly massless. However, the same effects that induce a potential for the axion also induce an axion mass. Indeed, from Eq. (2.34) we immediately see that

$$m_a^2 \equiv \left.\frac{\partial^2 V}{\partial a^2}\right|_{a=f_a\theta_{\text{QCD}}} = \frac{\kappa^2}{f_a^2}\,,$$

(2.35)

This seems like a contradiction, because a mass term breaks the shift symmetry and for a true Nambu-Goldstone boson, we expect this term to vanish. However, in the presence of quark masses, the transformations in Eq. (2.30) do not leave the Lagrangian invariant under the shift symmetry

$$m_q \overline{\psi}_R \psi_L \to m_q \overline{\psi}'_R e^{2ia/f_a} \psi'_L\,.$$

(2.36)

Fermion masses lead to an explicit breaking of the shift symmetry and turn the axion a pseudo Nambu-Goldstone boson, similar to the pions in QCD. For more than one quark flavor it suffices to have a single massless quark to recover the shift symmetry. We can determine this mass term from a chiral Lagrangian in which the fundamental fields are hadrons instead of quarks and find

$$m_a^2 = \frac{m_u m_d}{(m_u + m_d)^2}\frac{m_\pi^2 f_\pi^2}{f_a^2} \approx \frac{m_\pi^2 f_\pi^2}{2 f_a^2}\,,$$

(2.37)

where $f_\pi \approx m_\pi \approx 140\,\text{MeV}$ are the pion decay constant and mass, respectively. This term vanishes in the limit $m_u \to 0$ or $m_d \to 0$ as we expect from the discussion above. In the original axion proposal, $f_a \sim v$ and therefore $m_a \sim 10\,\text{keV}$. Since the couplings of the axion are also fixed by the value of f_a, such a particle was excluded very fast by searches for rare kaon decays, like for instance $K^+ \to \pi^+ a$. In general, f_a can be a free parameter and the mass of the axion can be smaller and its couplings can be weaker.

This leaves the question for which model parameters the axion makes a good dark matter candidate. Since the value of the axion field is not necessarily at the minimum of the potential at the time of the QCD phase transition, the axion begins to oscillates around the minimum and the oscillation energy density contributes to the dark matter relic density. This is a special version of the more general misalignment mechanism described in the previous section. We can then employ Eq. (2.26) and find the relation for the observed relic density

$$m_a a(t_{\mathrm{prod}}) \approx (m_a M_{\mathrm{Pl}})^{3/4} \, \mathrm{eV}^{1/2} \,. \tag{2.38}$$

The maximum field value of the oscillation mode is given by $a(t_{\mathrm{prod}}) \approx f_a$ and therefore

$$m_a f_a \overset{(2.37)}{\approx} m_\pi f_\pi \qquad \Rightarrow \qquad \boxed{m_a \approx \frac{(m_\pi f_\pi)^{4/3}}{M_{\mathrm{Pl}}} \, \mathrm{eV}^{-2/3}} \,. \tag{2.39}$$

This relation holds for $m_a \approx 2 \cdot 10^{-6} \, \mathrm{eV}$, which corresponds to $f_a \approx 10^{13} \, \mathrm{GeV}$. For heavier axions and smaller values of f_a, the axion can still constitute a part of the relic density. For example with a mass of $m_a = 6 \cdot 10^{-5} \, \mathrm{eV}$ and $f_a \approx 3 \cdot 10^{11} \, \mathrm{GeV}$, axions make up one per-cent of the observed relic density.

Dark Matter candidates with such low masses are hard to detect and we usually takes advantage of their couplings to photons. In Eq. (2.31), there is no reason why the coupling c_γ needs to be there. It is neither relevant for the strong CP problem nor for the axion to be dark matter. However, from the perspective of the effective theory we expect all couplings which are allowed by the assumed symmetry structure to appear. This includes the axion coupling to photons. If the complete theory, the axion coupling to gluons needs to be induced by some physics at the mass scale f_a. This can be achieved by axion couplings to SM quarks, or by axion couplings to non-SM fields that are color-charged but electrically neutral. Even in the latter case there is a non-zero coupling induced by the axion mixing with the SM pion after the QCD phase transition. Apart from really fine-tuned models the axion therefore couples to photons with an order-one coupling constant c_γ.

In Fig. 2.1 the yellow band shows the range of axion couplings to photons for which the models solve Eq. (2.37). The regime where the axion is a viable dark matter candidate is dashed. It is notoriously hard to probe axion dark matter in the parameter space in which they can constitute dark matter. Helioscopes try to convert axions produced in the sun into observable photons through a strong magnetic field. Haloscopes like ADMX use the same strategy to search for axions in the dark matter halo.

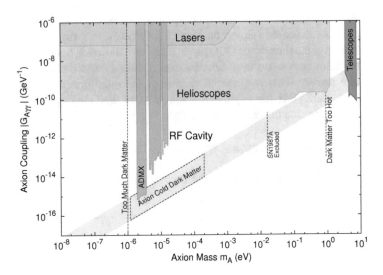

Fig. 2.1 Range of masses and couplings for which the axion can be a viable cold dark matter candidate. Figure from Ref. [1]

The same axion coupling to photons that we rely on for axion detection also allows for the decay $a \to \gamma\gamma$. This looks dangerous for a dark matter candidate, to we can estimate the corresponding decay width,

$$\Gamma(a \to \gamma\gamma) = \frac{\alpha^3}{256\,\pi^3} \frac{m_a^3}{f_a^2} \, |c_\gamma|^2 \tag{2.40}$$

$$= \frac{1}{137^3} \frac{1}{256\,\pi^3} \frac{(6 \cdot 10^{-6}\,\text{eV})^3}{(10^{22}\,\text{eV})^2} \, |c_\gamma|^2 \approx 1 \cdot 10^{-70}\,\text{eV}\,|c_\gamma|^2$$

Assuming $c_\gamma = 1$ this corresponds to a lifetime of $\tau = 1/\Gamma \approx 2 \cdot 10^{47}$ years, many orders of magnitude larger than the age of the universe.

While the axion is particularly interesting because it addresses the strong CP problem and dark matter at the same time, we can drop the relation to the CP problem and study axion-like particles (ALPs) as dark matter. For such a light pseudoscalar particle Eqs. (2.37) and (2.39) are replaced by the more general relations

$$m_a = \frac{\mu^2}{f_a} \quad \Rightarrow \quad m_a = \frac{\mu^{8/3}}{M_{\text{Pl}}}\,\text{eV}^{-2/3}\,. \tag{2.41}$$

where μ is a mass scale not related to QCD. In such models, the axion-like pseudoscalar can be very light. For example for $\mu \approx 100\,\text{eV}$, the axion mass is $m_a \approx 10^{-22}\,\text{eV}$. For such a low mass and a typical velocity of $v \approx 100\,\text{km/s}$, the de-Broglie wavelength is around 1 Kpc, the size of a galaxy. This type of dark matter is called fuzzy dark matter and can inherit the interesting properties of Bose-Einstein condensates or super-fluids.

2.4 Matter vs Anti-matter

Before we look at a set of relics linked to dark matter, let us follow a famous argument fixing the conditions which allow us to live in a Universe dominated by matter rather than anti-matter. In this section we will largely follow Kolb and Turner [2]. The observational reasoning why matter dominates the Universe goes in two steps: first, matter and anti-matter cannot be mixed, because we do not observe constant macroscopic annihilation; second, if we separate matter and anti-matter we should see a boundary with constant annihilation processes, which we do not see, either. So there cannot be too much anti-matter in the Universe.

The corresponding measurement is usually formulated in terms of the observed baryons, protons and neutrons, relative to the number of photons,

$$\frac{n_B}{n_\gamma} \approx 6 \cdot 10^{-10} . \tag{2.42}$$

The normalization to the photon density is motivated by the fact that this ratio should be of order unity in the very early Universe. Effects of the Universe's expansion and cooling to first approximation cancel. Its choice is only indirectly related to the observed number of photons and instead assumes that the photon density as the entropy density in thermal equilibrium. As a matter of fact, we use this number already in Eq. (1.53).

To understand Eq. (2.42) we start by remembering that in the hot universe anti-quarks and quarks or anti-baryons and baryons are pair-produced out of a thermal bath and annihilate with each other in thermal equilibrium. Following the same argument as for the photons, the baryons and anti-baryons decouple from each other when the temperature drops enough. In this scenario we can estimate the ratio of baryon and photon densities from Eq. (1.40), assuming for example $T = 20\,\mathrm{MeV} \ll m_B = 1\,\mathrm{GeV}$

$$\frac{n_B(T)}{n_\gamma(T)} = \frac{n_{\bar{B}}(T)}{n_\gamma(T)} = \frac{g_B \left(\frac{m_B T}{2\pi}\right)^{3/2} e^{-m_B/T}}{\frac{\zeta_3}{\pi^2} g_\gamma T^3}$$

$$= \frac{g_B}{g_\gamma} \frac{\sqrt{\pi}}{2\sqrt{2}\zeta_3} \left(\frac{m_B}{T}\right)^{3/2} e^{-m_B/T} = 3.5 \cdot 10^{-20} \tag{2.43}$$

The way of looking at the baryon asymmetry is that independent of the actual anti-baryon density the density of baryons observed today is much larger than what we would expect from thermal production. While we will see that for dark matter the problem is to get their interactions just right to produce the correct freeze-out density, for baryons the problem is to avoid their annihilation as much as possible.

We can think of two ways to avoid such an over-annihilation in our thermal history. First, there could be some kind of mechanism stopping the annihilation of baryons and anti-baryons when n_B/n_γ reaches the observed value. The problem with this solution is that we would still have to do something with the anti-baryons, as discussed above.

The second solution is to assume that through the baryon annihilation phase there exists an initially small asymmetry, such that almost all anti-baryons annihilate while the observed baryons remain. As a rough estimate, neglecting all degrees of freedom and differences between fermions and bosons, we assume that in the hot thermal bath we start with roughly as many baryons as photons. After cooling we assume that the anti-baryons reach their thermal density given in Eq. (2.43), while the baryons through some mechanism arrive at today's density given in Eq. (2.42). The baryon vs anti-baryon asymmetry starting at an early time then becomes

$$\frac{n_B - n_{\bar{B}}}{n_B} \approx \frac{n_B}{n_\gamma} - \frac{n_{\bar{B}}}{n_\gamma} \xrightarrow{\text{cooling}} 6 \cdot 10^{-10} - 3.5 \cdot 10^{-20} \approx 6 \cdot 10^{-10} . \qquad (2.44)$$

If we do the proper calculation, the correct number for a net quark excess in the early Universe comes out around

$$\boxed{\frac{n_B - n_{\bar{B}}}{n_B} \approx 3 \cdot 10^{-8}} . \qquad (2.45)$$

In the early Universe we start with this very small net asymmetry between the very large individual densities of baryons and anti-baryons. Rather than through the freeze-out mechanism introduced for neutrinos in Sect. 2.1, the baryons decouple when all anti-baryons are annihilated away. This mechanism can explain the very large baryon density measured today. The question is now how this asymmetry occurs at high temperatures.

Unlike the rest of the lecture notes, the discussion of the matter anti-matter asymmetry is not aimed at showing how the relic densities of the two species are computed. Instead, we will get to the general Sakharov conditions which tell us what ingredients our theory has to have to generate a net baryon excess in the early Universe, where we naively would expect the number of baryons and anti-baryons (or quarks and anti-quarks) to be exactly the same and in thermal equilibrium. Let us go through these condition one by one:

Baryon number violation—to understand this condition we just need to remember that we want to generate a different density of baryons (baryon number $B = +1$) and anti-baryons (baryon number $B = -1$) dynamically during the evolution of the Universe. We assume that our theory is described by a Lagrangian including finite temperature effects. If our Lagrangian is fully symmetric with respect to exchanging baryons with anti-baryons there will be no effect, no interaction, no scattering rate, no decay rate, nothing that can ever distinguish between baryons and anti-baryons and hence generate an asymmetry from a symmetric starting point. Let us assume that we want to generate a baryon asymmetry from an interaction of quarks and leptons with a heavy state X of the kind

$$X \to dd \qquad\qquad X \to \bar{d}\ell^- , \qquad\qquad (2.46)$$

where the d quark carries baryon number 1/3. A scattering process induced by these two interactions,

$$dd \to X^* \to \bar{d}\ell^- , \qquad\qquad (2.47)$$

links an initial state with $B = 2/3$ to a final state with $B = -1/3$. The combination $B - L$ is instead conserved. Such heavy bosons can appear in grand unified theories.

In the Standard Model the situation is a little more complicated: instead of the lepton number L and the baryon number B individually, the combination $B - L$ is indeed an (accidental) global symmetry of the electroweak interaction to all orders. In contrast, the orthogonal $B + L$ is anomalous, i.e. there are quantum contributions to scattering processes which respect $B - L$ but violate $B + L$. One can show that non-perturbative finite-temperature sphaleron processes can generate the combined state

$$\epsilon_{ijk} \left(u_{L,i} d_{L,j} u_{L,k} e_L + \cdots \right) \qquad\qquad (2.48)$$

for one generation of fermions with $SU(2)_L$ indices i, j, k out of the vacuum. It violates lepton and baryon number,

$$\Delta L = 1 \qquad\qquad \Delta B = 1 \qquad\qquad \Delta(B - L) = 0 . \qquad (2.49)$$

The probability of these sphaleron transition to happen at zero temperature (where they are called instanton transitions) scales like $e^{-8\pi^2/g^2}$ with the weak $SU(2)_L$ coupling $g \approx 0.7$. At high temperatures their rate increases significantly. The main effect of such interactions is that we can replace the condition of baryon number violation with lepton number violation when we ensure that sphaleron-induced processes transform a lepton asymmetry into a baryon asymmetry and neither of them gets washed out. This process is called leptogenesis rather than baryogenesis.

Departure from thermal equilibrium—in our above setup we can see what assumptions we need to be able to generate a net baryon asymmetry from the interactions given in Eq. (2.46) and the scattering given in Eq. (2.47). If we follow the reasoning for the relic photons we reduce the temperature until the two sides of the $2 \to 2$ scattering process in Eq. (2.47) drop out of thermal equilibrium. Our Universe could settles on one of the two sides of the scattering process, i.e. either with a net excess of d over $\bar{d}$ particles or vice versa. The problem is that the process $\bar{d}\bar{d} \to \bar{X}^* \to d\ell^+$ with $m_X = m_{\bar{X}}$ is protected by CPT invariance and will compensate everything exactly.

The more promising approach are out-of-equilibrium decays of the heavy X boson. This means that a population of X and $\bar{X}$ bosons decouple from the thermal bath early and induce the baryon asymmetry through late decays preferably into quarks or anti-quarks. In both cases we see that baryon number violating interactions require a departure from thermal equilibrium to generate a net baryon asymmetry in the evolution of the Universe.

In the absence of late-decaying particles, for example in the Standard Model, we need to rely on another mechanism to deviate from thermal equilibrium. The electroweak phase transition, like any phase transition, can proceed in two ways: if the phase transition is of first order the Higgs potential develops a non-trivial minimum while we are sitting at the unbroken field value $\phi = 0$. At the critical temperature the broken minimum becomes the global minimum of the potential and we have to tunnel there. The second order phase transition instead develops the broken minimum smoothly such that there is never a potential barrier between the two and we can smoothly transition into the broken minimum around the critical temperature. For a first-order phase transition different regions of the Universe will switch to the broken phase at different times, starting with expanding bubbles of broken phase regions. At the bubble surface the thermal equilibrium will be broken, allowing for a generation of the baryon asymmetry through the electroweak phase transition. Unfortunately, the Standard Model Higgs mass would have had to be below 60 GeV to allow for this scenario. C and CP violation—this condition appears more indirectly.

First, even if we assume that a transition of the kind shown in Eq. (2.46) exists we need to generate a baryon asymmetry from these decays where the heavy state and its anti-particle are produced from the vacuum. Charge conjugation links particles and anti-particles, which means that C conservation implies

$$\Gamma(X \to dd) = \Gamma(\bar{X} \to \bar{d}\bar{d}) \qquad \text{and} \qquad \Gamma(X \to d\ell^-) = \Gamma(\bar{X} \to \bar{d}\ell^+) \,. \tag{2.50}$$

In that case there will always be the same numbers of baryons d and anti-baryons $\bar{d}$ on average in the system. We only quote the statement that statistical fluctuations of the baryon and anti-baryon numbers are not large enough to explain the global asymmetry observed.

Next we assume a theory where C is violated, but CP is intact. This could for example be the electroweak Standard Model with no CP-violating phases in the quark and lepton mixing matrices. For our toy model we introduce a quark chirality $q_{L,R}$ which violates parity P but restores CP as a symmetry. For our decay widths transforming under C and CP this means

$$\Gamma(X \to d_L d_L) \neq \Gamma(\bar{X} \to \bar{d}_L \bar{d}_L) \qquad \qquad \text{(C violation)}$$

$$\Gamma(X \to d_L d_L) = \Gamma(\bar{X} \to \bar{d}_R \bar{d}_R) \qquad \qquad \text{(CP conservation)}$$

$$\Gamma(X \to d_R d_R) = \Gamma(\bar{X} \to \bar{d}_L \bar{d}_L) \qquad \text{(CP conservation)}$$

$$\Rightarrow \qquad \Gamma(X \to dd) \equiv \Gamma(X \to d_L d_L) + \Gamma(X \to d_R d_R)$$

$$= \Gamma(\bar{X} \to \bar{d}_R \bar{d}_R) + \Gamma(\bar{X} \to \bar{d}_L \bar{d}_L)$$

$$= \Gamma(\bar{X} \to \bar{d}\bar{d}) \tag{2.51}$$

This means unless C and CP are both violated, there will be no baryon asymmetry from X decays to d quarks.

In the above argument there is, strictly speaking, one piece missing: if we assume that we start with the same number of X and $\bar{X}$ bosons out of thermal equilibrium, once all of them have decayed to dd and $\bar{d}\bar{d}$ pairs irrespective of their chirality there is again no asymmetry between d and $\bar{d}$ quarks in the Universe. An asymmetry only occurs if a competing X decay channel produces a different number of baryons and allows the different partial widths to generate a net asymmetry. This is why we include the second term in Eq. (2.46). Assuming C and CP violation it implies

$$\Gamma(X \to dd) \neq \Gamma(\bar{X} \to \bar{d}\bar{d})$$

$$\Gamma(X \to \bar{d}\ell^-) \neq \Gamma(\bar{X} \to d\ell^+) \qquad \text{but} \quad \Gamma_{\text{tot}}(X) = \Gamma_{\text{tot}}(\bar{X}) \tag{2.52}$$

because of CPT invariance, just like $m_X = m_{\bar{X}}$.

2.5 Asymmetric Dark Matter

Starting from the similarity of the measured baryon and dark matter densities in Eq. (1.75)

$$\frac{\Omega_\chi}{\Omega_b} = \frac{0.12}{0.022} = 5.5 , \tag{2.53}$$

an obvious question is if we can link these two matter densities. We know that the observed baryon density in the Universe today is not determined by a thermal freeze-out, but by an initial small asymmetry between the baryon and anti-baryon densities. If we assume that dark matter is very roughly as heavy as baryons, that dark matter states carry some kind of charge which defines dark matter anti-particles, and that the baryon and dark matter asymmetries are linked, we can hope to explain the observed dark matter relic density. Following the leptogenesis example we could assume that the sphaleron transition not only breaks $B + L$, but also some kind of dark matter number D. Dark matter is then generated thermally, but the value of the relic density is not determined by thermal freeze-out. Still, from the structure formation constraints discussed in Sect. 1.5 we know that the dark matter agent should not be too light.

First, we can roughly estimate the dark matter masses this scenario predicts. From Sect. 2.4 we know how little we understand about the mechanism of generating the baryon asymmetry in models structurally similar to the Standard Model. For that reason, we start by just assuming that the particle densities of the baryons and of dark matter trace each other through some kind of mechanism,

$$n_\chi(T) \approx n_B(T) . \tag{2.54}$$

This will start in the relativistic regime and remain true after the two sectors decouple from each other and both densities get diluted through the expansion of the Universe. For the observed densities by PLANCK we use the non-relativistic relation between number and energy densities in Eqs. (1.40) and (1.41),

$$\frac{\Omega_\chi}{\Omega_b} = \frac{\rho_\chi}{\rho_B} = \frac{m_\chi n_\chi}{m_B n_B} \approx \frac{m_\chi}{m_B} \qquad \Leftrightarrow \qquad m_\chi \approx 5.5\, m_B \approx 5\, \text{GeV} . \tag{2.55}$$

Corrections to this relation can arise from the mechanism linking the two asymmetries.

Alternatively, we can assume that at the temperature T_{dec} at which the link between the baryons and the dark matter decouples, the baryons are relativistic and dark matter is non-relativistic. For the two energy densities this means

$$\rho_\chi(T_{\text{dec}}) = m_\chi n_\chi(T_{\text{dec}})$$

$$\approx m_\chi n_B(T_{\text{dec}}) = m_\chi \frac{30\zeta_3}{\pi^4} \frac{\rho_B(T_{\text{dec}})}{T_{\text{dec}}}$$

$$\Rightarrow \qquad \frac{m_\chi}{T_{\text{dec}}} = \frac{\rho_\chi(T_{\text{dec}})}{\rho_B(T_{\text{dec}})} \frac{\pi^4}{30\zeta_3} \approx 15 . \tag{2.56}$$

The relevant temperature is determined by the interactions between the baryonic and the dark matter sectors. However, in general this scenario will allow for heavy dark matter, $m_\chi \gg m_B$.

In a second step we can analyze what kind of dark matter annihilation rates are required in the asymmetric dark matter scenario. Very generally, we know the decoupling condition of a dark matter particle of the thermal bath of Standard Model states from the relativistic case. The mediating process can include a Standard Model fermion, $\chi f \to \chi f$. The corresponding annihilation process for dark matter which is not its own anti-particle is

$$\chi \bar{\chi} \to f \bar{f} . \tag{2.57}$$

As long as these scattering processes are active, the dark matter agent follows the decreasing temperature of the light Standard Model states in an equilibrium between production out of the thermal bath and annihilation. At some point, dark-matter freezes out of the thermal bath, and its density is only reduced by the expansion of the Universe. This point of decoupling is roughly given by Eq. (2.4), or

$$n_\chi(T_{dec}) = \frac{H}{\sigma_{\chi\chi} v} = \frac{\pi\sqrt{g_{eff}(T_{dec})}}{\sqrt{90}M_{Pl}} \frac{T_{dec}^2}{\sigma_{\chi\chi} v} \tag{2.58}$$

in terms of the dark matter annihilation cross section $\sigma_{\chi\chi}$.

The special feature of asymmetric dark matter is that this relation does not predict the dark matter density $n_\chi(T_{dec})$ leading to the observed relic density. Instead, this annihilation has to remove all dark matter anti-particles and almost all dark matter particles, while the observed relic density is generated by a very small dark matter asymmetry. If we follow the numerical example of the baryon asymmetry given in Eq. (2.45) this means we need a dark matter annihilation rate which is 10^8 times the rate necessary to predict the observed relic density for pure freeze-out dark matter. From the typical expressions for cross sections in Eq. (1.49) we see that a massless mediator or a t-channel diagram in combination with light dark matter leads to large cross sections,

$$\sigma_{\chi\chi} \approx \frac{\pi\alpha_\chi^2}{m_\chi^2}, \tag{2.59}$$

with the generic dark matter coupling α_χ to a dark gauge boson or another light mediator. For heavier dark matter we will see in Sect. 4.1 how we can achieve large annihilation rates through a $2 \to 1$ annihilation topology.

References

1. Tanabashi, M., et al.: [Particle data group], Review of particle physics. Phys. Rev. D **98**(3), 030001 (2018)
2. Kolb, E.W., Turner, M.S.: The early Universe. Front. Phys. **69**, 1 (1990)

Chapter 3
Thermal Relic Density

After introducing the observed relic density of photons in Sect. 1.3 and the observed relic density of neutrinos in Sect. 2.1 we will now compute the relic density of a hypothetical massive, weakly interacting dark matter agent. As for the photons and neutrinos we assume dark matter to be created thermally, and the observed relic density to be determined by the freeze-out combined with the following expansion of the Universe. We will focus on masses of at least a few GeV, which guarantees that dark matter will be non-relativistic when it decouples from thermal equilibrium. At this point we do not have specific particles in mind, but in Chap. 4 we will illustrate this scenario with a set of particle physics models.

The general theme of this chapter and the following Chaps. 5–7 is the typical four-point interaction of the dark matter agent with the Standard Model. For illustration purposes we assume the dark matter agent to be a fermion χ and the Standard Model interaction partner a fermion f:

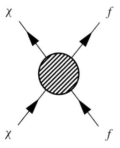

Unlike for asymmetric dark matter, in this process it does not matter if the dark matter agent has an anti-particle $\bar{\chi}$, or if is it's own anti-particle $\chi = \bar{\chi}$.

© Springer Nature Switzerland AG 2019
M. Bauer, T. Plehn, *Yet Another Introduction to Dark Matter*,
Lecture Notes in Physics 959, https://doi.org/10.1007/978-3-030-16234-4_3

This Feynman diagram, or more precisely this amplitude mediates three different scattering processes:

- left-to-right we can compute dark matter annihilation, $\chi \bar{\chi} \rightarrow f\bar{f}$, see this chapter and Chaps. 4 and 5;
- bottom-to-top it describes dark matter scattering of visible matter $\chi f \rightarrow \chi f$, see Chap. 6;
- right-to-left it describes dark matter pair-production, $f\bar{f} \rightarrow \chi \bar{\chi}$, see Chap. 7.

This strong link between very different observables is what makes dark matter so interesting for particle physicists, including the possibility of global analyses for any model which can predict this amplitude. Note also that we will see how different the kinematics of the different scattering processes actually are.

3.1   WIMP Miracle

As for the relativistic neutrinos, we will first avoid solving the full Boltzmann equation for the number density as a function of time. Instead, we assume that some kind of interaction keeps the dark matter particle χ in thermal equilibrium with the Standard Model particles and at the same time able to annihilate. At the point of thermal decoupling the dark matter freezes out with a specific density. As for the neutrinos, the underlying process is described by the matrix element for dark matter annihilation

$$\chi \chi \rightarrow f\bar{f} \,. \tag{3.1}$$

As in Eq. (1.51) the interaction rate Γ corresponding to this scattering process just compensates the increasing scale factor at the point of decoupling,

$$\Gamma(T_{\text{dec}}) \overset{!}{=} H(T_{\text{dec}}) \,. \tag{3.2}$$

Assuming this interaction rate is set by electroweak interactions, for non-relativistic dark matter agents, the temperature dependence in Eq. (1.49) vanishes and gets replaced by the dark matter mass. To allow for an s-channel process in Eq. (3.1) we use the Z-mass and Z-coupling in the corresponding annihilation cross section

$$\boxed{\sigma_{\chi\chi}(T \ll m_\chi) = \frac{\pi \alpha^2 m_\chi^2}{c_w^4 m_Z^4}} \,. \tag{3.3}$$

This formula combines the dark matter mass m_χ with a weak interaction represented by a $1/m_Z$ suppression, implicitly assuming $m_\chi \ll m_Z$. We will check this assumption later. Following Eq. (2.2) we can use the non-relativistic number density. For the non-relativistic decoupling we should not assume $v = 1$, as we did before.

Given the limited number of energy scales in our description we instead estimate very roughly

$$\frac{m_\chi}{2}v^2 = T \qquad \Leftrightarrow \qquad v = \sqrt{\frac{2T}{m_\chi}} \,, \tag{3.4}$$

remembering that we need to check this later. Moreover, we set the number of relevant degrees of freedom of the dark matter agent to $g = 2$, corresponding for example to a complex scalar or a Majorana fermion. In that case the condition of dark matter freeze-out is

$$\Gamma := \sigma_{\chi\chi} v n_\chi \stackrel{\text{Eq.(1.40)}}{=} \sigma_{\chi\chi} \sqrt{\frac{2T_{\text{dec}}}{m_\chi}} \, g \left(\frac{m_\chi T_{\text{dec}}}{2\pi}\right)^{3/2} e^{-m_\chi/T_{\text{dec}}}$$

$$\stackrel{!}{=} H \stackrel{\text{Eq.(1.47)}}{=} \frac{\pi}{3\sqrt{10}\,M_{\text{Pl}}} \sqrt{g_{\text{eff}}(T_{\text{dec}})} \, T_{\text{dec}}^2 \,. \tag{3.5}$$

We can solve this condition and find

$$\sigma_{\chi\chi} \frac{m_\chi T_{\text{dec}}^2}{\pi^{3/2}} e^{-x_{\text{dec}}} = \frac{\pi}{3\sqrt{10}\,M_{\text{Pl}}} \sqrt{g_{\text{eff}}(T_{\text{dec}})} T_{\text{dec}}^2 \qquad \text{with} \qquad \boxed{x := \frac{m_\chi}{T}}$$

$$\Leftrightarrow \qquad e^{-x_{\text{dec}}} = \frac{\pi^{5/2}}{3\sqrt{10}\,m_\chi M_{\text{Pl}}\,\sigma_{\chi\chi}} \frac{\sqrt{g_{\text{eff}}(T_{\text{dec}})}}{}$$

$$= 1.8 \frac{\sqrt{g_{\text{eff}}(T_{\text{dec}})}}{m_\chi M_{\text{Pl}}\,\sigma_{\chi\chi}} \,. \tag{3.6}$$

Note how in this calculation the explicit temperature dependence drops out. This means the result can be considered an equation for the ratio x_{dec}. If we want to include the temperature dependence of g_{eff} we cannot solve this equation in a closed form, but we can estimate the value of x_{dec}. First, we can use the generic electroweak annihilation cross section from Eq. (3.3) to find

$$e^{-x_{\text{dec}}} = \frac{\pi\sqrt{\pi}}{3\sqrt{10}\,\alpha^2} \frac{c_w^4 m_Z^4}{m_\chi^3 M_{\text{Pl}}} \sqrt{g_{\text{eff}}(T_{\text{dec}})} \,. \tag{3.7}$$

Next, we assume that most of the Standard Model particles contribute to the active degrees of freedom. From Eq. (1.45) we know that the full number gives us $g_{\text{eff}} = 106.75$. In the slightly lower range $T_{\text{dec}} = 5 \ldots 80\,\text{GeV}$ the weak bosons and the top quark decouple, and Eq. (1.44) gives the slightly reduced value

$$g_{\text{eff}}(T_{\text{dec}}) = (8 \times 2 + 2) + \frac{7}{8} (5 \times 3 \times 2 \times 2 + 3 \times 2 \times 2 + 3 \times 2)$$

$$= 18 + \frac{7}{8} 78 = 86.25 \,. \tag{3.8}$$

Combining all prefactors we find the range

$$
e^{-x_{\text{dec}}} \approx 6 \cdot 10^5 \, \frac{m_Z^4}{m_\chi^3 \, M_{\text{Pl}}} = \begin{cases} 2 \cdot 10^{-9} & \leftrightarrow & x_{\text{dec}} \approx 20 & (m_\chi = 10\,\text{GeV}) \\ 6 \cdot 10^{-11} & \leftrightarrow & x_{\text{dec}} \approx 23 & (m_\chi = 30\,\text{GeV}) \\ 8 \cdot 10^{-12} & \leftrightarrow & x_{\text{dec}} \approx 26 & (m_\chi = 60\,\text{GeV}) . \end{cases}
$$
(3.9)

As a benchmark we will use $m_\chi = 30\,\text{GeV}$ with $x_{\text{dec}} \approx 23$ from now on. We need to eventually check these assumptions, but because of the leading exponential dependence we expect this result for x_{dec} to be insensitive to our detailed assumptions. Following Eqs. (1.40) and (3.7) the temperature at the point of decoupling gives us the non-relativistic number density at the point of decoupling,

$$
n_\chi(T_{\text{dec}}) = g \left(\frac{m_\chi T_{\text{dec}}}{2\pi} \right)^{3/2} e^{-x_{\text{dec}}} = \frac{\pi}{3\sqrt{20}\, M_{\text{Pl}}} \sqrt{\frac{m_\chi}{T_{\text{dec}}}} \sqrt{g_{\text{eff}}(T_{\text{dec}})} \, T_{\text{dec}}^2 \, \frac{c_w^4 m_Z^4}{\pi \alpha^2 m_\chi^2}
$$

$$
\approx 10^3 \, \frac{m_Z^4}{M_{\text{Pl}}} \left(\frac{T_{\text{dec}}}{m_\chi} \right)^{3/2} \approx \frac{10^3}{x_{\text{dec}}^{3/2}} \, \frac{m_Z^4}{M_{\text{Pl}}} .
$$
(3.10)

From the time of non-relativistic decoupling we have to evolve the energy density to the current time or temperature T_0. We start with the fact that once a particle has decoupled, its number density drops like $1/a^3$, as we can read off Eq. (1.27) in the non-relativistic case,

$$
\rho_\chi(T_0) = m_\chi \, n_\chi(T_0) = m_\chi \, n_\chi(T_{\text{dec}}) \left(\frac{a(T_{\text{dec}})}{a(T_0)} \right)^3 .
$$
(3.11)

To translate this dependence on the scale factor a into a temperature dependence we need to quote the same, single thermodynamic result as in Sect. 2.2, namely that according to Eq. (1.33) the combination $a(T)\,T$ is almost constant. When we take into account the active degrees of freedom and their individual temperature dependence the relation is more precisely

$$
\left(\frac{a(T_{\text{dec}})T_{\text{dec}}}{a(T_0)T_0} \right)^3 = \frac{g_{\text{eff}}(T_0)}{g_{\text{eff}}(T_{\text{dec}})} \approx \frac{3.6}{100} = \frac{1}{28} ,
$$
(3.12)

again for $T_{\text{dec}} > 5\,\text{GeV}$ and depending slightly on the number of neutrinos we take into account. We can use this result to compute the non-relativistic energy density now

$$
\rho_\chi(T_0) = m_\chi \left(\frac{a(T_{\text{dec}})T_{\text{dec}}}{a(T_0)T_0} \right)^3 \frac{T_0^3}{T_{\text{dec}}^3} \, n_\chi(T_{\text{dec}}) = \frac{x_{\text{dec}}}{28} \, T_0^3 \, \frac{n_\chi(T_{\text{dec}})}{T_{\text{dec}}^2}
$$

$$
= T_0^3 \, \frac{n_\chi(T_{\text{dec}})x_{\text{dec}}^3}{28 m_\chi^2} \overset{\text{Eq. (3.10)}}{\approx} 3 \cdot 10^3 \, \frac{m_Z^4}{m_\chi^2 M_{\text{Pl}}} \, T_0^3 .
$$
(3.13)

Using this result we can compute the dimensionless dark matter density in close analogy to the neutrino case of Eq. (2.11),

$$\Omega_\chi h^2 = \frac{\rho_\chi(T_0)h^2}{3M_{Pl}^2 H_0^2}$$

$$\approx 3 \cdot 10^3 \frac{m_Z^4}{m_\chi^2 M_{Pl}} \frac{(2.4 \cdot 10^{-4})^3}{(2.5 \cdot 10^{-3})^4} \frac{h^2}{\text{eV}}$$

$$\approx 3 \cdot 10^3 \frac{7 \cdot 10^7}{2 \cdot 10^{18}} \frac{\text{GeV}^3}{m_\chi^2} \frac{1}{5} \frac{10^9}{\text{GeV}} \approx 20 \frac{\text{GeV}^2}{m_\chi^2}$$

$$\Leftrightarrow \qquad \boxed{\Omega_\chi h^2 \approx 0.12 \left(\frac{13\,\text{GeV}}{m_\chi}\right)^2}. \qquad (3.14)$$

This outcome is usually referred to as the WIMP miracle: if we assume an dark matter agent with an electroweak-scale mass and an annihilation process mediated by the weak interaction, the predicted relic density comes out exactly as measured.

Let us recapitulate where the WIMP mass dependence of Eq. (3.14) comes from: first, the annihilation cross section in Eq. (3.3) is assumed to be mediated by electroweak interactions and includes a dependence on m_χ. Our original assumption $m_\chi \ll m_W$ is not perfectly fulfilled, but also not completely wrong. Second, the WIMP mass enters the relation between the number and energy density, but some of this dependence is absorbed into the value $x_{dec} = 23$, which means that the decoupling of the non-relativistic WIMPs is supposed to happen at a very low temperature of $T_{dec} \approx m_\chi/23$. Making things worse, some of the assumption we made in this non-relativistic and hence multi-scale calculation are not as convincing as they were for the simpler relativistic neutrino counterpart, so let us check Eq. (3.14) with an alternative estimate. One of the key questions we will try to answer in our alternative approach is how the m_χ-dependence of Eq. (3.14) occurs.

3.2 Boltzmann Equation

Because the derivation for the non-relativistic dark matter agent is at the heart of these lecture notes, we will also show how to properly compute the current relic density of a weakly interacting, massive dark matter agent. This calculation is based on the Boltzmann equation. It describes the change of a number density $n(t)$ with time. The first effect included in the equation is the increasing scale factor $a(t)$. It even occurs in full equilibrium,

$$0 = \frac{d}{dt}\left[n(t)a(t)^3\right] = \dot{n}(t)a(t)^3 + 3n(t)a(t)^2\dot{a}(t)$$

$$\Leftrightarrow \dot{n}(t) + 3H(t)n(t) = 0. \qquad (3.15)$$

At some point, the underlying assumption of thermal equilibrium breaks down. For the number of WIMPs the relevant process is not a process which guarantees thermal equilibrium with other states, but the explicit pair production or pair annihilation via a weakly interacting process

$$\chi\chi \leftrightarrow f\bar{f} \,, \tag{3.16}$$

with any available pair of light fermions in the final state. The depletion rate from the WIMP pair annihilation process in Eq. (3.16) is given by the corresponding $\sigma_{\chi\chi}\, v\, n_\chi^2$. This rate describes the probability of the WIMP annihilation process in Eq. (3.16) to happen, given the WIMP density and their velocity. For the relativistic relic neutrinos we could safely assume $v = 1$, while for the WIMP case we did not even make this assumption for our previous order-of-magnitude estimate.

When we derive the Boltzmann equation from first principles it turns out that we need to thermally average. This reflects the fact that the WIMP number density is a global observable, integrated over the velocity spectrum. In the non-relativistic limit the velocity of a particle with momentum $\vec{k}$ and energy k_0 is

$$v_k := \frac{|\vec{k}|}{k_0} \approx \frac{|\vec{k}|}{m_\chi} \ll 1 \,. \tag{3.17}$$

The external momenta of the two fermions then have the form

$$k^2 = k_0^2 - \vec{k}^2 = k_0^2 - (m_\chi v_k)^2 \overset{!}{=} m_\chi^2$$

$$\Leftrightarrow k_0 = \sqrt{m_\chi^2 + m_\chi^2 v_k^2} \approx m_\chi \left(1 + \frac{v_k^2}{2}\right) \,. \tag{3.18}$$

For a $2 \rightarrow 2$ scattering process we have to distinguish the velocities of the individual states and the relative velocity. The energy of initial state is given by the Mandelstam variable $s = (k_1 + k_2)^2$, in terms of the incoming momenta k_1 and k_2. These momenta are linked to the masses of the incoming dark matter state via $k_1^2 = k_2^2 = m_\chi^2$. For two incoming states with the same mass this gives us the velocity of each of the two particles as

$$s = (k_1 + k_2)^2 = 2m_\chi^2 + 2k_1^0 k_2^0 - 2\vec{k}_1\vec{k}_2 \overset{\text{cms}}{=} 2m_\chi^2 + 2(k_1^0)^2 + 2|\vec{k}_1|^2$$

$$= 4m_\chi^2 + 4|\vec{k}_1|^2 = 4m_\chi^2(1 + v_1^2) \quad \Leftrightarrow \quad v_1^2 = \frac{s - 4m_\chi^2}{4m_\chi^2} = \frac{s}{4m_\chi^2} - 1 \,. \tag{3.19}$$

The relative velocity of the two incoming particles in the non-relativistic limit is instead defined as

$$v = \left| \frac{\vec{k}_1}{k_1^0} - \frac{\vec{k}_2}{k_2^0} \right|_{\text{cms}} \stackrel{}{=} \left| \frac{\vec{k}_1}{k_1^0} + \frac{\vec{k}_1}{k_1^0} \right| = \frac{2|\vec{k}_1|}{k_1^0} \approx 2v_1$$

$$\Leftrightarrow m_\chi^2 v^2 = 4 m_\chi^2 v_1^2 = s - 4 m_\chi^2 \ . \tag{3.20}$$

Using the relative velocity the thermal average of $\sigma_{\chi\chi} v$ as it for example appears in Eq. (3.7) is defined as

$$\langle \sigma_{\chi\chi \to ff} \, v \rangle := \frac{\int d^3 p_{\chi,1} d^3 p_{\chi,2} \, e^{-(E_{\chi,1}+E_{\chi,2})/T} \, \sigma_{\chi\chi \to ff} \, v}{\int d^3 p_{\chi,1} d^3 p_{\chi,2} \, e^{-(E_{\chi,1}+E_{\chi,2})/T}}$$

$$= \frac{2\pi^2 T \int_{4m_\chi^2}^{\infty} ds \, \sqrt{s}(s - 4m_\chi^2) K_1 \left(\frac{\sqrt{s}}{T} \right) \sigma_{\chi\chi \to ff}(s)}{\left(4\pi m_\chi^2 T \, K_2 \left(\frac{m_\chi}{T} \right) \right)^2} \ , \tag{3.21}$$

in terms of the modified Bessel functions of the second kind $K_{1,2}$. Unfortunately, this form is numerically not very helpful in the general case. The thermal averaging replaces the global value of $\sigma_{\chi\chi} \, v$, as it gets added to the equilibrium Boltzmann equation (Eq. (3.15)) on the right-hand side,

$$\boxed{\dot{n}(t) + 3H(t)n(t) = -\langle \sigma_{\chi\chi} \, v \rangle \left(n(t)^2 - n_{\text{eq}}(t)^2 \right) .} \tag{3.22}$$

The time dependence of n induced by the annihilation process is proportional to n^2 because of the two WIMPs in the initial state of the annihilation process. The form of the equation guarantees that for $n = n_{\text{eq}}$ the only change in density occurs from the expanding Universe.

We can analytically solve this Boltzmann equation using a set of approximations. We start with a re-definition, introducing the yield Y, we get rid of the linear term,

$$\frac{1}{a(t)^3} \frac{d}{dt} \left(n(t)a(t)^3 \right) = -\langle \sigma_{\chi\chi} \, v \rangle \left(n(t)^2 - n_{\text{eq}}(t)^2 \right)$$

$$\Leftrightarrow \quad T(t)^3 \frac{d}{dt} \left(\frac{n(t)}{T(t)^3} \right) = -\langle \sigma_{\chi\chi} \, v \rangle \left(n(t)^2 - n_{\text{eq}}(t)^2 \right)$$

$$\Leftrightarrow \quad \frac{dY(t)}{dt} = -\langle \sigma_{\chi\chi} \, v \rangle \, T(t)^3 \left(Y(t)^2 - Y_{\text{eq}}(t)^2 \right)$$

$$\text{with} \quad Y(t) := \frac{n(t)}{T^3} \ . \tag{3.23}$$

Throughout these lecture notes we have always replaced the time by some other variable describing the history of the Universe. We again switch variables to $x = m_\chi/T$. For the Jacobian we assume that most of the dark matter decoupling happens with $\rho_r \gg \rho_m$; in the early, radiation-dominated Universe we can link the time and x through the Hubble constant,

$$\frac{1}{2t} \overset{\text{Eq. (1.31)}}{=} H \overset{\text{Eq. (1.47)}}{=} \frac{H(x=1)}{x^2} \qquad \Leftrightarrow \qquad x = \sqrt{2tH(x=1)}$$

$$\Leftrightarrow \qquad \frac{dx}{dt} = \frac{2H(x=1)}{2\sqrt{2tH(x=1)}} = \frac{H(x=1)}{x} ,$$
$$(3.24)$$

where $x = 1$ means $T = m_\chi$. With this Jacobian the Boltzmann equation becomes

$$\frac{dY(x)}{dx} = \frac{x}{H(x=1)} \frac{dY(t)}{dt}$$

$$= -\langle \sigma_{\chi\chi} v \rangle \frac{x}{H(x=1)} \frac{m_\chi^3}{x^3} \left(Y(x)^2 - Y_{\text{eq}}(x)^2 \right)$$

$$= -\frac{\lambda(x)}{x^2} \left(Y(x)^2 - Y_{\text{eq}}(x)^2 \right)$$

$$\text{with} \quad \lambda(x) := \frac{m_\chi^3 \langle \sigma_{\chi\chi} v \rangle}{H(x=1)} = \frac{\sqrt{90} \, M_{\text{Pl}} m_\chi}{\pi \sqrt{g_{\text{eff}}}} \langle \sigma_{\chi\chi} v \rangle(x) . \qquad (3.25)$$

To analytically solve this Boltzmann equation we make two approximations: first, according to Eq. (1.40) the equilibrium density drops like e^{-x} towards later times or increasing x. Assuming that the actual number density $n(x)$ drops more slowly, we can safely approximate the Boltzmann equation by

$$\frac{dY(x)}{dx} = -\frac{\lambda(x)}{x^2} Y(x)^2 . \qquad (3.26)$$

Second, we can estimate $\lambda(x)$ by expanding the thermally averaged annihilation WIMP cross section for small velocities. We use Eq. (3.3) as the leading term in the annihilation cross section and approximate v following Eq. (3.4), giving us

$$\lambda(x) = \frac{\sqrt{90} \, M_{\text{Pl}} m_\chi}{\pi \sqrt{g_{\text{eff}}}} \sigma_{\chi\chi} v + \mathcal{O}(v^2)$$

$$\approx \frac{\sqrt{90} \, M_{\text{Pl}} m_\chi}{\pi \sqrt{g_{\text{eff}}}} \sqrt{\frac{2}{x}} \frac{\pi \alpha^2 m_\chi^2}{c_w^4 m_Z^4} \equiv \frac{\bar{\lambda}}{\sqrt{x}} . \qquad (3.27)$$

The value $\bar{\lambda}$ depends on x independently through g_{eff}, so we can assume it to be constant as long as g_{eff} does not change much. Under this assumption we can then solve the Boltzmann equation with the simple substitution $\overline{Y} = 1/Y$,

$$\frac{dY(x)}{dx} = \frac{d}{dx}\frac{1}{\overline{Y}} = -\frac{1}{\overline{Y}(x)^2}\frac{d\overline{Y}(x)}{dx} \overset{!}{=} -\frac{\bar{\lambda}}{x^{5/2}}\frac{1}{\overline{Y}(x)^2} \quad \Leftrightarrow \quad \frac{d\overline{Y}(x)}{dx} = \frac{\bar{\lambda}}{x^{5/2}} .$$

$$(3.28)$$

From Eq. (3.9) we know that thermal WIMPs have masses well above $10\,\text{GeV}$, which corresponds to $g_{\text{eff}} \approx 100$. This value only changes once the temperature reaches the bottom mass and then drops to $g_{\text{eff}} \approx 3.6$ today. This allows us to separate the leading effects driving the dark matter density into the decoupling phase described by the Boltzmann equation and an expansion phase with its drop in g_{eff}. For the first phase we can just integrate the Boltzmann equation for constant g_{eff} starting just before decoupling (x_{dec}) and to a point $x'_{\text{dec}} \gg x_{\text{dec}}$ after decoupling but above the bottom mass,

$$\frac{1}{Y(x'_{\text{dec}})} - \frac{1}{Y(x_{\text{dec}})} = \overline{Y}(x'_{\text{dec}}) - \overline{Y}(x_{\text{dec}}) = -\frac{\bar{\lambda}}{x'^{3/2}_{\text{dec}}} + \frac{\bar{\lambda}}{x^{3/2}_{\text{dec}}} .$$

$$(3.29)$$

From the form of the Boltzmann equation in Eq. (3.26) we see that $Y(x)$ drops rapidly with increasing x. If we choose $x'_{\text{dec}} \gg x_{\text{dec}} = 23$ it follows that $Y(x'_{\text{dec}}) \ll Y(x_{\text{dec}})$ and hence

$$\frac{1}{Y(x'_{\text{dec}})} = \frac{\bar{\lambda}}{x^{3/2}_{\text{dec}}}$$

$$Y(x'_{\text{dec}}) = \frac{m^3_\chi \langle \sigma_{\chi\chi} v \rangle}{H(x=1)} = \frac{x_{\text{dec}}}{\lambda(x_{\text{dec}})} \overset{\text{Eq. (3.25)}}{=} x_{\text{dec}} \frac{\pi\sqrt{g_{\text{eff}}}}{\sqrt{90}\,M_{\text{Pl}}m_\chi} \frac{1}{\langle \sigma_{\chi\chi} v \rangle} .$$

$$(3.30)$$

In this expression g_{eff} is evaluated around the point of decoupling. For the second, expansion phase we can just follow Eq. (3.11) and compute

$$\rho_\chi(T_0) = m_\chi n_\chi(T_0)$$

$$= m_\chi Y(x'_{\text{dec}}) T'^3_{\text{dec}} \left(\frac{a(T'_{\text{dec}})}{a(T_0)}\right)^3 \overset{\text{Eq. (3.12)}}{=} m_\chi Y(x'_{\text{dec}}) T^3_0 \frac{g_{\text{eff}}(T_0)}{g_{\text{eff}}(T'_{\text{dec}})}$$

$$= m_\chi \frac{Y(x'_{\text{dec}})T^3_0}{28} .$$

$$(3.31)$$

For the properly normalized relic density this means

$$\Rightarrow \quad \Omega_\chi h^2 = m_\chi \, \frac{Y(x'_{\text{dec}}) T_0^3}{28} \, \frac{h^2}{3 M_{\text{Pl}}^2 H_0^2}$$

$$= \frac{h^2 \pi \sqrt{g_{\text{eff}}}}{28\sqrt{90}\, M_{\text{Pl}}} \, \frac{x_{\text{dec}}}{\langle \sigma_{\chi\chi} v \rangle} \, \frac{T_0^3}{3 M_{\text{Pl}}^2 H_0^2}$$

$$= \frac{h^2 \pi \sqrt{g_{\text{eff}}}}{28\sqrt{90}\, M_{\text{Pl}}} \, \frac{x_{\text{dec}}}{\langle \sigma_{\chi\chi} v \rangle} \, \frac{(2.4 \cdot 10^{-4})^3}{(2.5 \cdot 10^{-3})^4} \, \frac{1}{\text{eV}} \tag{3.32}$$

$$\Rightarrow \quad \boxed{\Omega_\chi h^2 \approx 0.12 \, \frac{x_{\text{dec}}}{23} \, \frac{\sqrt{g_{\text{eff}}}}{10} \, \frac{1.7 \cdot 10^{-9}\, \text{GeV}^{-2}}{\langle \sigma_{\chi\chi} v \rangle}} \, .$$

We can translate this result into different units. In the cosmology literature people often use $\text{eV}^{-1} = 2 \cdot 10^{-5}$ cm. In particle physics we measure cross sections in barn, where $1\,\text{fb} = 10^{-39}\,\text{cm}^2$. Our above result is a very good approximation to the correct value for the relic density in terms of the annihilation cross section

$$\Omega_\chi h^2 \approx 0.12 \, \frac{x_{\text{dec}}}{23} \, \frac{\sqrt{g_{\text{eff}}}}{10} \, \frac{1.7 \cdot 10^{-9}\, \text{GeV}^{-2}}{\langle \sigma_{\chi\chi} v \rangle}$$

$$\approx 0.12 \, \frac{x_{\text{dec}}}{23} \, \frac{\sqrt{g_{\text{eff}}}}{10} \, \frac{2.04 \cdot 10^{-26}\,\text{cm}^3/\text{s}}{\langle \sigma_{\chi\chi} v \rangle} \, . \tag{3.33}$$

With this result we can now insert the WIMP annihilation rate given by Eqs. (3.3) and (3.4),

$$\langle \sigma_{\chi\chi} v \rangle = \sigma_{\chi\chi} v + \mathcal{O}(v^2) \approx \sqrt{\frac{2}{x}} \, \frac{\pi \alpha^2 m_\chi^2}{c_w^4 m_Z^4}$$

$$\Rightarrow \quad \Omega_\chi h^2 = 0.12 \, \frac{x_{\text{dec}}}{23} \, \frac{\sqrt{g_{\text{eff}}}}{10} \, \frac{c_w^4 m_Z^4 \sqrt{x}}{\sqrt{2}\pi \alpha^2 m_\chi^2} \, \frac{1.7 \cdot 10^{-9}}{\text{GeV}^2}$$

$$= 0.12 \, \left(\frac{x_{\text{dec}}}{23}\right)^{3/2} \frac{\sqrt{g_{\text{eff}}}}{10} \, \left(\frac{35\,\text{GeV}}{m_\chi}\right)^2 \, . \tag{3.34}$$

We can compare this result to our earlier estimate in Eq. (3.14) and confirm that these numbers make sense for a weakly interacting particle with a weak-scale mass.

Alternatively, we can replace the scaling of the annihilation cross section given in Eq. (3.3) by a simpler form, only including the WIMP mass and certainly valid for heavy dark matter, $m_\chi > m_Z$. We find

$$\boxed{\langle \sigma_{\chi\chi} v \rangle \approx \frac{g^4}{16\pi m_\chi^2} \overset{!}{=} \frac{1.7 \cdot 10^{-9}}{\text{GeV}^2}} \qquad \Leftrightarrow \qquad g^2 \approx \frac{m_\chi}{3400\,\text{GeV}} = \frac{m_\chi}{3.4\,\text{TeV}}.$$

$$(3.35)$$

This form of the cross section does not assume a weakly interacting origin, it simply follows for the scaling with the coupling and from dimensional analysis. Depending on the coupling, its prediction for the dark matter mass can be significantly higher. Based on this relation we can estimate an upper limit on m_χ from the unitarity condition for the annihilation cross section

$$g^2 \lesssim 4\pi \qquad \Leftrightarrow \qquad m_\chi < 54\,\text{TeV} . \qquad (3.36)$$

A lower limit does not exist, because we can make a lighter particle more and more weakly coupled. Eventually, it will be light enough to be relativistic at the point of decoupling, bringing us back to the relic neutrinos discussed in Sect. 2.1.

Let us briefly recapitulate our argument which through the Boltzmann equation leads us to the WIMP miracle: we start with a $2 \to 2$ scattering process linking dark matter to Standard Model particles through a so-called mediator, which can for example be a weak boson. This allows us to compute the dark matter relic density as a function of the mediating coupling and the dark matter mass, and it turns out that a weak-coupling combined with a dark matter mass below the TeV scale fits perfectly. There are two ways in which we can modify the assumed dark matter annihilation process given in Eq. (3.16): first, in the next section we will introduce additional annihilation channels for an extended dark matter sector. Second, in Sect. 4.1 we will show what happens if the annihilation process proceeds through an s-channel Higgs resonance.

3.3 Co-annihilation

In many models the dark matter sector consists of more than one particle, separated from the Standard Model particles for example through a specific quantum number. A typical structure are two dark matter particles χ_1 and χ_2 with $m_{\chi_1} < m_{\chi_2}$. In analogy to Eq. (3.16) they can annihilate into a pair of Standard Model particles through the set of processes

$$\chi_1\chi_1 \to f\bar{f} \qquad \chi_1\chi_2 \to f\bar{f} \qquad \chi_2\chi_2 \to f\bar{f} . \qquad (3.37)$$

This set of processes can mediate a much more efficient annihilation of the dark matter state χ_1 together with the second state χ_2, even in the limit where the actual dark matter process $\chi_1\chi_1 \to f\bar{f}$ is not allowed. Two non-relativistic states will have number densities both given by Eq. (1.40). We know from Eq. (3.9) that decoupling of a WIMP happens at typical values $x_{dec} = m_\chi / T_{dec} \approx 28$, so if we for example assume $\Delta m_\chi = m_{\chi_2} - m_{\chi_1} = 0.2\, m_{\chi_1}$ and $g_1 = g_2$ we find

$$\frac{n_2(T_{dec})}{n_1(T_{dec})} \overset{\text{Eq.(1.40)}}{=} \frac{g_2}{g_1} \left(1 + \frac{\Delta m_\chi}{m_{\chi_1}}\right)^{3/2} e^{-\Delta m_\chi / T_{dec}} = 1.31\, e^{-0.2 x_{dec}} \approx \frac{1}{206}.$$
$$(3.38)$$

Just from statistics the heavier state will already be rare by the time the lighter, actual dark matter agent annihilates. For a mass difference around 10% this suppression is reduced to a factor 1/15, gives us an estimate that efficient co-annihilation will prefer two states with mass differences in the 10% range or closer.

Let us assume that there are two particles present at the time of decoupling. In addition, we assume that the first two processes shown in Eq. (3.37) contribute to the annihilation of the dark matter state χ_1. In this case the Boltzmann equation from Eq. (3.22) reads

$$\dot{n}_1(t) + 3H(t)n_1(t)$$

$$= -\langle \sigma_{\chi_1 \chi_1} v \rangle \left(n_1(t)^2 - n_{1,eq}(t)^2\right) - \langle \sigma_{\chi_1 \chi_2} v \rangle \left(n_1(t)n_2(t) - n_{1,eq}(t)n_{2,eq}(t)\right)$$

$$\approx -\langle \sigma_{\chi_1 \chi_1} v \rangle \left(n_1(t)^2 - n_{1,eq}(t)^2\right) - \langle \sigma_{\chi_1 \chi_2} v \rangle \left(n_1^2(t) - n_{1,eq}(t)^2\right) \frac{n_2}{n_1}$$

$$\overset{\text{Eq.(3.38)}}{=} - \left[\langle \sigma_{\chi_1 \chi_1} v \rangle + \langle \sigma_{\chi_1 \chi_2} v \rangle \frac{g_2}{g_1} \left(1 + \frac{\Delta m_\chi}{m_{\chi_1}}\right)^{3/2} e^{-\Delta m_\chi / T} \right]$$

$$\times \left(n_1(t)^2 - n_{1,eq}(t)^2\right).$$
$$(3.39)$$

In the second step we assume that the two particles decouple simultaneously, such that their number densities track each other through the entire process, including the assumed equilibrium values. This means that we can throughout our single-species calculations just replace

$$\langle \sigma_{\chi\chi} v \rangle \rightarrow \langle \sigma_{\chi_1 \chi_1} v \rangle + \langle \sigma_{\chi_1 \chi_2} v \rangle \frac{g_2}{g_1} \left(1 + \frac{\Delta m_\chi}{m_{\chi_1}}\right)^{3/2} e^{-\Delta m_\chi / T}.$$
$$(3.40)$$

In the co-annihilation setup it is not required that the direct annihilation process dominates. The annihilation of more than one particle contributing to a dark matter sector can include many other aspects, for example when the dark matter state only interacts gravitationally and the annihilation proceeds mostly through a next-to-lightest, weakly interacting state. The Boltzmann equation will in this case split into one equation for each state and include decays of the heavier state into the dark matter state. Such a system of Boltzmann equations cannot be solved analytically in general.

What we can assume is that the two co-annihilation partners have very similar masses, $\Delta m_\chi \ll m_{\chi_1}$, similar couplings, $g_1 = g_2$, and that the two annihilation processes in Eq. (3.37) are of similar size, $\langle \sigma_{\chi_1\chi_1} v \rangle \approx \langle \sigma_{\chi_1\chi_2} v \rangle$. In that limit we simply find $\langle \sigma_{\chi\chi} v \rangle \rightarrow 2\langle \sigma_{\chi_1\chi_1} v \rangle$ in the Boltzmann equation. We know from Eq. (3.33) how the correct relic density depends on the annihilation cross section. Keeping the relic density constant we absorb the rate increase through co-annihilation into a shift in the typical WIMP masses of the two dark matter states. According to Eq. (3.35) the WIMP masses should now be

$$\langle \sigma_{\chi\chi} v \rangle \approx \frac{g^4}{16\pi m_\chi^2} \equiv 2 \frac{g^4}{32\pi m_{\chi_1}^2} \qquad \text{or} \qquad m_{\chi_1} \approx m_{\chi_2} \approx \sqrt{2} m_\chi . \tag{3.41}$$

A simple question we can ask for example when we will talk about collider signatures is how easy it would be to discover a single WIMP compared to the pair of co-annihilating, slightly heavier WIMPs.

An interesting question is how co-annihilation channels modify the WIMP mass scale which is required by the observed relic density. From Eq. (3.41) we see that an increase in the total annihilation rate leads to a larger mass scale of the dark matter particles, as expected from our usual scaling. On the other hand, the annihilation cross section really enters for example Eq. (3.33) in the combination $\langle \sigma_{\chi_1\chi_1} v \rangle / \sqrt{g_{\text{eff}}}$. If we increase the number of effective degrees of freedom significantly, while the co-annihilation channels really have a small effect on the total annihilation rate, the dark matter mass might also decrease.

3.4 Velocity Dependence

While throughout the early estimates we use the dark matter annihilation rate $\sigma_{\chi\chi}$, we introduce the more appropriate thermal expectation value of the velocity times the annihilation rate $\langle \sigma_{\chi\chi} v \rangle$ in Eq. (3.21). This combination has the nice feature that its leading term can be independent of the velocity v. In general, the velocity-weighted cross section will be of the form

$$\langle \sigma_{\chi\chi} v \rangle = \langle s_0 + s_1 v^2 + \mathcal{O}(v^4) \rangle \tag{3.42}$$

This pattern follows from the partial wave analysis of relativistic scattering. The first term s_0 is velocity-independent and arises from S-wave scattering. An example is the scattering of two scalar dark matter particles with an s-channel scalar mediator or two Dirac fermions with an s-channel vector mediator. The second term s_1 with a vanishing rate at threshold is generated by S-wave and P-wave scattering. It occurs for example for Dirac fermion scattering through an s-channel vector mediator. All t-channel processes have an S-wave component and are not suppressed at threshold.

For dark matter phenomenology, the dependence on a potentially small velocity shown in Eq. (3.42) is the important aspect. Different dark matter agents with different interaction patterns lead to distinct threshold dependences. For s-channel and t-channel mediators and several kinds of couplings to Standard Model fermions f in the final state we find

	s-Channel mediator				t-Channel mediator			
	$\bar{f}f$	$\bar{f}\gamma^5 f$	$\bar{f}\gamma^\mu f$	$\bar{f}\gamma^\mu\gamma^5 f$	$\bar{f}f$	$\bar{f}\gamma^5 f$	$\bar{f}\gamma^\mu f$	$\bar{f}\gamma^\mu\gamma^5 f$
Dirac fermion	v^2	v^0	v^0	v^0	v^0	v^0	v^0	v^0
Majorana fermion	v^2	v^0	0	v^0	v^0	v^0	v^0	v^0
Real scalar	v^0	v^0	0	0				
Complex scalar	v^0	v^0	v^2	v^2				

Particles who are their own anti-particles, like Majorana fermions and real scalars, do not annihilate through s-channel vector mediators. The same happens for complex scalars and axial-vector mediators. In general, t-channel annihilation to two Standard Model fermions is not possible for scalar dark matter.

To allow for an efficient dark matter annihilation to today's relic density, we tend to prefer an un-suppressed contribution s_0 to increase the thermal freeze-out cross section. The problem with such large annihilation rates is that they are strongly constrained by early-universe physics. For example, the PLANCK measurements of the matter power spectrum discussed in Sect. 1.5 constrain the light dark matter very generally, just based on the fact that such light dark matter can affect the photon background at the time of decoupling. The problem arises if dark matter candidates annihilate into Standard Model particles through non-gravitational interactions,

$$\chi\chi \to SM\,SM \ . \tag{3.43}$$

As we know from Eq. (3.1) this process is the key ingredient to thermal freeze-out dark matter. If it happens at the time of last scattering it injects heat into the intergalactic medium. This ionizes the hydrogen and helium atoms formed during recombination. While the ionization energy does not modify the time of the last scattering, it prolongs the period of recombination or, alternatively, leads to a broadening of the surface of last scattering. This leads to a suppression of the temperature fluctuations and enhance the polarization power spectrum. The temperature and polarization data from PLANCK puts an upper limit on the dark matter annihilation cross section

$$f_{\text{eff}}\frac{\langle\sigma_{\chi\chi}v\rangle}{m_\chi} \lesssim \frac{8.5\cdot 10^{-11}}{\text{GeV}^3} \tag{3.44}$$

The factor $f_{\text{eff}} < 1$ denotes the fraction of the dark matter rest mass energy injected into the intergalactic medium. It is a function of the dark matter mass, the dominant annihilation channel, and the fragmentation patterns of the SM particles the dark matter agents annihilate into. For example, a 200 GeV dark matter particle annihilating to photons or electrons reaches $f_{\text{eff}} = 0.66 \dots 0.71$, while an annihilation to muon pairs only gives $f_{\text{eff}} = 0.28$. As we know from Eq. (3.32) for freeze-out dark matter an annihilation cross section of the order $\langle \sigma_{\chi\chi} v \rangle \approx 1.7 \cdot 10^{-9} \, \text{GeV}^2$ is needed. This means that the PLANCK constraints of Eq. (3.44) requires

$$m_\chi \gtrsim 10 \, \text{GeV} , \qquad (3.45)$$

In contrast to limits from searches for dark matter annihilation in the center of the galaxy or in dwarf galaxies, as we will discuss in Chap. 5, this constraint does not suffer from astrophysical uncertainties, such as the density profile of the dark matter halo in galaxies.

3.5 Sommerfeld Enhancement

Radiative corrections can drastically change the threshold behavior shown in Eq. (3.42). As an example, we study the annihilation of two dark matter fermions through an s-channel scalar in the limit of small relative velocity of the two fermions. The starting point of our discussion is the loop diagram which describes the exchange of a gauge boson between two incoming (or outgoing) massive fermions χ:

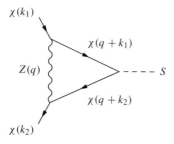

After inserting the Feynman rules we find the expression

$$\int d^4 q \, \frac{\slashed{q} + \slashed{k}_1 + m_\chi}{(q + k_1)^2 - m_\chi^2} \gamma_\mu \frac{1}{q^2 - m_Z^2} \gamma^\mu \frac{\slashed{q} + \slashed{k}_2 + m_\chi}{(q + k_2)^2 - m_\chi^2} . \qquad (3.46)$$

The question is where this integral receives large contributions. Using $k^2 = m_\chi^2$ the denominators of the fermion propagators read

$$\frac{1}{(q+k)^2 - m_\chi^2} = \frac{1}{q_0^2 - |\vec{q}|^2 + 2q_0 k_0 - 2\vec{q}\vec{k}}$$

$$\stackrel{\text{Eq. (3.18)}}{=} \frac{1}{q_0^2 - |\vec{q}|^2 + (2+v^2)m_\chi q_0 - 2m_\chi v|\vec{q}|\cos\theta + \mathcal{O}(q_0 v^2)}$$

$$\stackrel{|\vec{q}|=m_\chi v}{=} \frac{1}{q_0^2 - m_\chi^2 v^2(1 + 2\cos\theta) + (2+v^2)m_\chi q^0 + \mathcal{O}(q_0 v^2)} .$$

$$(3.47)$$

The particles in the loop are not on their respective mass shells. Instead, we can identify a particularly dangerous region for $v \to 0$, namely $q_0 = m_\chi v^2$, where

$$\frac{1}{(q+k)^2 - m_\chi^2} = \frac{1}{m_\chi^2 v^2(1 - 2\cos\theta) + \mathcal{O}(v^4)} . \tag{3.48}$$

Unless we make an assumption about the angle θ we cannot make a stronger statement about the contributions of the fermion propagators. If we just set $\cos\theta = 0$ we find

$$\frac{1}{(q+k)^2 - m_\chi^2} = \frac{1}{m_\chi^2 v^2 + \mathcal{O}(v^4)} . \tag{3.49}$$

In the same phase space region the Z boson propagator in the integral scales like

$$\frac{1}{q^2 - m_Z^2} = \frac{1}{m_\chi^2 v^4 - m_\chi^2 v^2 - m_Z^2} = -\frac{1}{m_\chi^2 v^2 + m_Z^2 + \mathcal{O}(v^4)} . \tag{3.50}$$

In the absence of the gauge boson mass the gauge boson propagator would diverge for $v \to 0$, just like the fermion propagators. This means that we can approximate the loop integral by focussing on the phase space regime

$$q_0 \approx m_\chi v^2 \qquad \text{and} \qquad |\vec{q}| \approx m_\chi v . \tag{3.51}$$

The complete infrared contribution to the one-loop matrix element of Eq. (3.46) with a massive gauge boson exchange and neglecting the Dirac matrix structure is

$$\int d^4q \, \frac{m_\chi}{(q+k_1)^2 - m_\chi^2} \frac{1}{q^2 - m_Z^2} \frac{m_\chi}{(q-k_2)^2 - m_\chi^2}$$

$$\approx \Delta q_0 (\Delta|\vec{q}|)^3 \frac{1}{m_\chi v^2} \frac{1}{m_\chi^2 v^2 + m_Z^2} \frac{1}{m_\chi v^2}$$

$$\approx m_\chi v^2 \, (m_\chi v)^3 \, \frac{1}{m_\chi v^2} \, \frac{1}{m_\chi^2 v^2 + m_Z^2} \, \frac{1}{m_\chi v^2}$$

$$= \frac{v}{v^2 + \dfrac{m_Z^2}{m_\chi^2}} \xrightarrow{m_\chi \gg m_Z} \frac{1}{v} \; . \tag{3.52}$$

This means that part of the one-loop correction to the dark matter annihilation process at threshold scales like $1/v$ in the limit of massless gauge boson exchange. For massive gauge bosons the divergent behavior is cut off with a lower limit $v \gtrsim m_Z/m_\chi$. If we attach an additional gauge boson exchange to form a two-loop integral, the above considerations apply again, but only to the last, triangular diagram. The divergence still has the form $1/v$. Eventually, it will be cut off by the widths of the particles, which is a phrase often used in the literature and not at all easy to show in detail.

What is more important is the question what the impact of this result is for our calculations—it will turn out that while the loop corrections for slowly moving particles with a massless gauge boson exchange are divergent, they typically correct a cross section which vanishes at threshold and only lead to a finite rate at the production threshold.

As long as we limit ourselves to $v \ll 1$ we do not need to use relativistic quantum field theory for this calculation. We can compute the same v-dependent correction to particle scattering using non-relativistic quantum mechanics. We assume two electrically and weakly charged particles $\chi^\pm$, so their attractive potential has spherically symmetric Coulomb and Yukawa parts,

$$V(r) = -\frac{e^2}{r} - \frac{g_Z^2}{r} e^{-m_Z r} \quad \text{with} \quad r = |\vec{r}| \; . \tag{3.53}$$

The coupling g_Z describes an unknown χ-χ-Z interaction. With such a potential we can compute a two-body scattering process. The wave function $\psi_k(\vec{r})$ will in general be a superposition of an incoming plane wave in the z-direction and a set of spherical waves with a modulation in terms of the scattering angle θ. As in Eq. (1.59) we can expand the wave function in spherical harmonics, combined with an energy-dependent radial function $R(r; E)$. We again exploit the symmetry with respect to the azimuthal angle ϕ and obtain

$$\psi_k(\vec{r}) = \sum_{\ell=0}^{\infty} \sum_{m=-\ell}^{\ell} a_{\ell m} Y_{\ell m}(\theta, \phi) \, R_\ell(r; E)$$

$$= \sum_{\ell=0}^{\infty} (2\ell + 1) \, a_{\ell 0} Y_{\ell 0}(\theta, \phi) \, R_\ell(r; E)$$

$$\overset{\text{Eq.(1.63)}}{=} \sum_{\ell=0}^{\infty} (2\ell+1)\, a_{\ell 0}\, \frac{\sqrt{2\ell+1}}{2}\, P_\ell(\cos\theta)\, R_\ell(r; E)$$

$$=: \sum_{\ell=0}^{\infty} A_\ell\, P_\ell(\cos\theta)\, R_\ell(r; E)\,. \tag{3.54}$$

From the calculation of the hydrogen atom we know that the radial, time-independent Schrödinger equation in terms of the reduced mass m reads

$$\left[-\frac{1}{2mr^2} \frac{d}{dr}\left(r^2 \frac{d}{dr}\right) + \frac{\ell(\ell+1)}{2mr^2} + V(r) - E \right] R_\ell(r; E) = 0\,. \tag{3.55}$$

The reduced mass for a system with two identical masses is given by

$$m = \frac{m_1 m_2}{m_1 + m_2} = \frac{m_\chi}{2}\,. \tag{3.56}$$

As a first step we solve the Schrödinger equation at large distances, where we can neglect $V(r)$. We know that the solution will be plane waves, but to establish our procedure we follow the procedure starting with Eq. (3.55) step by step,

$$\left[-\frac{1}{r^2}\frac{d}{dr}\left(r^2\frac{d}{dr}\right) + \frac{\ell(\ell+1)}{r^2} - k^2 \right] R_{k\ell}(r) = 0 \quad \text{with } k^2 := 2mE = m^2 v^2$$

$$\Leftrightarrow \frac{1}{\rho^2}\frac{d}{d\rho}\left(\rho^2\frac{dR_{k\ell}}{d\rho}\right) - \frac{\ell(\ell+1)}{\rho^2} R_{k\ell} + R_{k\ell} = 0 \quad \text{with } \rho := kr$$

$$\Leftrightarrow \frac{1}{\rho^2}\left(2\rho\frac{dR_{k\ell}}{d\rho} + \rho^2\frac{d^2R_{k\ell}}{d\rho^2}\right) - \frac{\ell(\ell+1)}{\rho^2} R_{k\ell} + R_{k\ell} = 0$$

$$\Leftrightarrow \rho^2\frac{d^2R_{k\ell}}{d\rho^2} + 2\rho\frac{dR_{k\ell}}{d\rho} - \ell(\ell+1)R_{k\ell} + \rho^2 R_{k\ell} = 0 \tag{3.57}$$

This differential equation turns out to be identical to the implicit definition of the spherical Bessel functions $j_\ell(\rho)$, so we can identify $R_{k\ell}(r) = j_\ell(\rho)$. The radial wave function can then be expressed in Legendre polynomials,

$$R_{k\ell}(r) = j_\ell(\rho) = \frac{1}{2\ell!}\left(\frac{\rho}{2}\right)^\ell (-1)^\ell \int_{-1}^{1} dt\, e^{i\rho t}\,(t^2-1)^\ell$$

$$= \frac{1}{2\ell!}\left(\frac{\rho}{2}\right)^\ell (-1)^\ell \left[\frac{1}{i\rho}e^{i\rho t}(t^2-1)^\ell \bigg|_{-1}^{1} \right.$$

$$\left. - \int_{-1}^{1} dt\, \frac{1}{i\rho}e^{i\rho t}\frac{d}{dt}(t^2-1)^\ell \right]$$

$$= \frac{1}{2\ell!} \left(\frac{\rho}{2}\right)^\ell (-1)^\ell \frac{(-1)}{i\rho} \int_{-1}^{1} dt\, e^{i\rho t} \frac{d}{dt} (t^2 - 1)^\ell = \cdots$$

$$= \frac{1}{2\ell!} \left(\frac{\rho}{2}\right)^\ell \frac{1}{(i\rho)^\ell} \int_{-1}^{1} dt\, e^{i\rho t} \frac{d^\ell}{dt^\ell} (t^2 - 1)^\ell$$

$$\overset{\text{Eq. (1.64)}}{=} \frac{(-i)^\ell}{2} \int_{-1}^{1} dt\, e^{i\rho t} P_\ell(t) . \tag{3.58}$$

The integration variable t corresponds to $\cos\theta$ in our physics problem. As mentioned above, these solutions to the free Schrödinger equation have to be plane waves. We use the relation

$$\sum_{\ell=0}^{\infty} \frac{2\ell + 1}{2} P_\ell(t) P_\ell(t') = \delta(t - t') \tag{3.59}$$

to link the plane wave to this expression in terms of the spherical Bessel functions and the Legendre polynomials. With the correct ansatz we find

$$\sum_{\ell=0}^{\infty} i^\ell (2\ell + 1) P_\ell(t) j_\ell(\rho) \overset{\text{Eq. (3.58)}}{=} \sum_{\ell=0}^{\infty} i^\ell (2\ell + 1) P_\ell(t) \frac{(-i)^\ell}{2} \int_{-1}^{1} dt'\, e^{i\rho t'} P_\ell(t')$$

$$\overset{\text{Eq. (3.59)}}{=} 2 i^\ell \frac{(-i)^\ell}{2} e^{i\rho t} = e^{ikr \cos\theta} . \tag{3.60}$$

If we know that the series in Eq. (3.54) describes such plane waves, we can determine $A_\ell R_{k\ell}$ by comparing the two sums and find

$$A_\ell\, R_{k\ell}(r) = i^\ell (2\ell + 1) j_\ell(kr) \approx \begin{cases} i^\ell (2\ell + 1) \dfrac{\sin\left(kr - \dfrac{\ell\pi}{2}\right)}{kr} & \text{for } kr \gg \ell^2 \\[4ex] i^\ell (2\ell + 1) \dfrac{(kr)^\ell}{(2\ell + 1)!!} & \text{for } kr \ll 2\sqrt{\ell} . \end{cases} \tag{3.61}$$

We include two limits which can be derived for the spherical Bessel functions. To describe the interaction with and without a potential $V(r)$ we are always interested in the wave function at the origin. The lower of the above two limits indicates that for small r and hence small ρ values only the first term $\ell = 0$ will contribute. We can evaluate $j_0(kr)$ for $kr = 0$ in both forms and find the same value,

$$\left|\psi_k(\vec{0})\right|^2 \overset{\ell=0}{=} |A_0 P_0(\cos\theta) R_{k\ell}(0)|^2 = |A_0 R_{k\ell}(0)|^2 = \lim_{r\to 0} |j_0(kr)|^2 = 1 \tag{3.62}$$

The argument that only $\ell = 0$ contributes to the wave function at the origin is not at all trivial to make, and it holds as long as the potential does not diverge faster than $1/r$ towards the origin.

Next, we add an attractive Coulomb potential to Eq. (3.57), giving us the radial Schrödinger equation in a slightly re-written form in the first term

$$\left[-\frac{1}{r}\frac{d^2}{dr^2}r + \frac{\ell(\ell+1)}{r^2} - \frac{2me^2}{r} - k^2 \right]\frac{u_{k\ell}}{r} = 0 \quad \text{with } u_{k\ell}(r) := r R_{k\ell}(r)$$

$$\Leftrightarrow \frac{d^2}{dr^2}u_{k\ell} - \frac{\ell(\ell+1)}{r^2}u_{k\ell} + \frac{2me^2}{r}u_{k\ell} + k^2 u_{k\ell} = 0$$

$$\Leftrightarrow \frac{d^2}{d\rho^2}u_{k\ell} - \frac{\ell(\ell+1)}{\rho^2}u_{k\ell} + \frac{2me^2}{\rho k}u_{k\ell} + u_{k\ell} = 0 \tag{3.63}$$

The solution of this equation will lead us to the well-known hydrogen atom and its energy levels. However, we are not interested in the energy levels but in the continuum scattering process. Following the discussion around Eq. (3.61) and assuming that the Coulomb potential will not change the fundamental structure of the solution around the origin we can evaluate the radial wave function for $\ell = 0$,

$$\frac{d^2}{d\rho^2}u_{k0} + \frac{2me^2}{\rho k}u_{k0} + u_{k0} = 0 \tag{3.64}$$

This is the equation we need to solve and then evaluate at the origin, $\vec{r} = \vec{0}$. We only quote the result,

$$\left| \psi_k(\vec{0}) \right|^2 = \frac{2\pi e^2}{v}\frac{1}{1 - e^{-2\pi e^2/v}} \approx \begin{cases} \dfrac{2\pi e^2}{v} & \text{for } v \to 0 \\[2mm] 1 & \text{for } v \to \infty \end{cases}. \tag{3.65}$$

Compared to Eq. (3.62) this increased probability measure is called the Sommerfeld enhancement. It is divergent at small velocities, just as in the Feynman-diagrammatic discussion before. For very small velocities, it can lead to an enhancement of the threshold cross section by several orders of magnitude.

It can be shown that the calculation based on ladder diagrams in momentum space and based on the Schrödinger equation in position space are equivalent for simple scattering processes. The resummation of the ladder diagrams is equivalent to the computation of the wave function at the origin in the Fourier-transformed position space.

The case of the Yukawa potential shows a similar behavior. It involves an amusing trick in the computation of the potential, so we discuss it in some detail. When we include the Yukawa potential in the Schrödinger equation we cannot solve the equation analytically; however, the Hulthen potential is an approximation to the Yukawa potential which does allow us to solve the Schrödinger equation. It is defined as

$$V(r) = \frac{g_Z^2 \delta e^{-\delta r}}{1 - e^{-\delta r}} . \tag{3.66}$$

Optimizing the numerical agreement of the Hulthen potential's radial wave functions with those of the Yukawa potential suggests for the relevant mass ratio in our calculation

$$\delta \approx \frac{\pi^2}{6} m_Z , \tag{3.67}$$

which we will use later. Unlike for the Coulomb potential we can now keep the full ℓ-dependence of the Schrödinger equation. The only additional approximation we use is for the angular momentum term

$$\frac{\delta^2 e^{-\delta r}}{\left(1 - e^{-\delta r}\right)^2} = \delta^2 \frac{1 - \delta r + \mathcal{O}(\delta^2 r^2)}{\left(-\delta r + \frac{1}{2}\delta^2 r^2 + \mathcal{O}(\delta^3 r^3)\right)^2}$$

$$= \frac{1}{r^2} \frac{1 - \delta r + \mathcal{O}(\delta^2 r^2)}{\left(1 - \frac{1}{2}\delta r + \mathcal{O}(\delta^2 r^2)\right)^2} = \frac{1}{r^2}\left(1 + \mathcal{O}(\delta^2 r^2)\right) . \tag{3.68}$$

The radial Schrödinger equation of Eq. (3.63) with the Hulthen potential and the above approximation for the angular-momentum-induced potential term now reads

$$\left[-\frac{1}{r}\frac{d^2}{dr^2}r + \ell(\ell+1)\frac{\delta^2 e^{-\delta r}}{\left(1-e^{-\delta r}\right)^2} + \frac{g_Z^2 \delta e^{-\delta r}}{1-e^{-\delta r}} - k^2 \right]\frac{u_{k\ell}}{r} = 0$$

$$\Leftrightarrow \quad \frac{d^2}{dr^2}u_{k\ell} - \ell(\ell+1)\frac{\delta^2 e^{-\delta r}}{\left(1-e^{-\delta r}\right)^2}u_{k\ell} + \frac{g_Z^2 \delta e^{-\delta r}}{1-e^{-\delta r}}u_{k\ell} + k^2 u_{k\ell} = 0 . \tag{3.69}$$

Again, we only quote the result: the leading term for the corresponding Sommerfeld enhancement factor in the limit $v \ll 1$ arises from

$$\left|\psi_k(\vec{0})\right|^2 = \frac{\pi g_Z^2}{v} \frac{\sinh\dfrac{2vm_\chi \pi}{\delta}}{\cosh\dfrac{2vm_\chi \pi}{\delta} - \cos\left(2\pi\sqrt{\dfrac{g_Z^2 m_\chi}{\delta} - \dfrac{v^2 m_\chi^2}{\delta^2}}\right)} . \tag{3.70}$$

This Sommerfeld enhancement factor will be a combination of a slowly varying underlying function with a peak structure defined by the denominator.

We are interested in the position and the height of the first and the following peaks. We need to two Taylor series

$$\sinh x = x + \mathcal{O}(x^3) \quad \text{and} \quad \cosh x = 1 + \frac{x^2}{2} + \mathcal{O}(x^4) \tag{3.71}$$

The cosh function is always larger than one and grows rapidly with increasing argument. This means that in the limit $v \ll 1$ the two terms in the denominator can cancel almost entirely,

$$\left| \psi_k(\vec{0}) \right|^2 = \frac{\pi g_Z^2}{v} \; \frac{\dfrac{2\pi v m_\chi}{\delta} + \mathcal{O}(v^3)}{1 + \mathcal{O}(v^2) - \cos\left(2\pi \sqrt{\dfrac{g_Z^2 m_\chi}{\delta}} + \mathcal{O}(v^2) \right)}$$

$$\xrightarrow{v \to 0} \; \frac{\dfrac{2\pi^2 g_Z^2 m_\chi}{\delta}}{1 - \cos\sqrt{\dfrac{4\pi^2 g_Z^2 m_\chi}{\delta}}} \; . \tag{3.72}$$

The finite limit for $v \to 0$ is well defined except for mass ratios m_χ/δ or m_χ/m_Z right on the pole. The positions of the peaks in this oscillating function of the mass ratio m_χ/m_Z is independent of the velocity in the limit $v \ll 1$. The peak positions are

$$\frac{4\pi^2 g_Z^2 m_\chi}{\delta} = (2n\pi)^2 \quad \Leftrightarrow \quad \frac{m_\chi}{\delta} = \frac{n^2}{g_Z^2}$$

$$\overset{\text{Eq.(3.67)}}{\Leftrightarrow} \quad \frac{m_\chi}{m_Z} = \frac{\pi^2}{6 g_Z^2} n^2 \quad \text{with } n = 1, 2, \dots \tag{3.73}$$

For example assuming $g_Z^2 \approx 1/20$ we expect the first peak at dark matter masses below roughly 3 TeV. For the Sommerfeld enhancement factor on the first peak we have to include the second term in the Taylor series in Eq. (3.70) and find

$$\left| \psi_k(\vec{0}) \right|^2 = \frac{\dfrac{2\pi^2 g_Z^2 m_\chi}{\delta}}{\dfrac{1}{2} \left(\dfrac{2 v m_\chi \pi}{\delta} \right)^2 + \mathcal{O}(v^4)} = \frac{g_Z^2 \delta}{m_\chi v^2}$$

$$\overset{\text{Eq.(3.73)}}{\Rightarrow} \quad \boxed{\left| \psi_k(\vec{0}) \right|^2 = \frac{g_Z^4}{v^2}} \; . \tag{3.74}$$

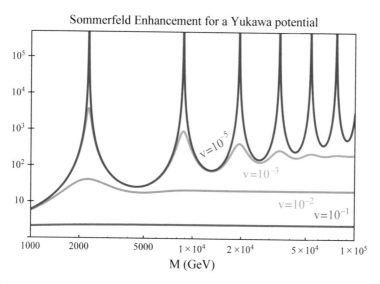

Fig. 3.1 Sommerfeld enhancement for a Yukawa potential as a function of the dark matter mass ($M \equiv m_\chi$), shown for different velocities. It assumes the correct Z-mass and a coupling strength of $g_Z^2 = 1/30$. Figure from Ref. [1], found for example in Mariangela Lisanti's lecture notes [2]

For $v = 10^{-3}$ and $g_Z^2 \approx 1/20$ we find sizeable Sommerfeld enhancement on the first peak by a factor around 2500. Figure 3.1 illustrates these peaks in the Sommerfeld enhancement for different velocities. The slightly different numerical values arise because the agreement of the Hulthen and Yukawa potentials is limited.

From our calculation and this final result it is clear that a large ratio of the dark matter mass to the electroweak masses modifies the pure v-dependence of the Coulomb-like Sommerfeld enhancement, but is not its source. Just like for the Coulomb potential the driving force behind the Sommerfeld enhancement is the vanishing velocity, leading to long-lived bound states. The ratio m_χ/m_Z entering the Sommerfeld enhancement is simply the effect of the Z-mass acting as a regulator towards small velocities.

3.6 Freeze-In Production

In the previous discussion we have seen that thermal freeze-out offers an elegant explanation of the observed relic density, requiring only minimal modifications to the thermal history of the Universe. On the other hand, for cold dark matter and asymmetric dark matter we have seen that an alternative production mechanism has a huge effect on dark matter physics. A crucial assumption behind freeze-out dark matter is that the coupling between the Standard Model and dark matter cannot be too small, otherwise we will never reach thermal equilibrium and cannot apply

Eq. (3.2). For example for the Higgs portal model discussed in Sect. 4.1 this is the case for a portal coupling of $\lambda_3 \lesssim 10^{-7}$. For such small interaction rates the (almost) model-independent lower bound on the dark matter mass from measurements of the CMB temperature variation and polarization, discussed in Sect. 1.4 and giving $m_\chi \gtrsim 10\,\text{GeV}$, does not apply. This allows for new kinds of light dark matter.

For such very weakly interacting particles, called feebly interacting massive particles or FIMPs, we can invoke the non-thermal, so-called freeze-in mechanism. The idea is that the dark matter sector gets populated through decay or annihilation of SM particles until the number density of the corresponding SM particles species becomes Boltzmann-suppressed. For an example SM particle B with an interaction

$$\mathcal{L} \ni -y m_B \bar{\chi} B \chi + h.c. \tag{3.75}$$

and $m_B > 2m_\chi$, the decay $B \rightarrow \chi\bar{\chi}$ allows to populate the dark sector. The Boltzmann equation in Eq. (3.22) then acquires a source term

$$\dot{n}_\chi(t) + 3H(t)n_\chi(t) = S(B \rightarrow \chi\bar{\chi}) . \tag{3.76}$$

The condition that the dark matter sector is not in thermal equilibrium initially translates into a lower bound on the dark matter mass. Its precise value depends on the model, but for a mediator with $m_B \approx 100\,\text{GeV}$ one can estimate $m_\chi \gtrsim 0.1 \ldots 1\,\text{keV}$ from the fundamental assumptions of the model.

A decay-based source term in terms of the internal number of degrees of freedom g_B^*, the partial width $B \rightarrow \chi\bar{\chi}$, and the equilibrium distribution $\exp(-E_B/T)$ can be written as

$$
\begin{aligned}
S(B &\rightarrow \chi\bar{\chi}) \\
&= g_B^* \int \frac{d^3 p_B}{(2\pi)^3} e^{-E_B/T} \frac{m_B}{E_B} \Gamma(B \rightarrow \chi\bar{\chi}) \\
&= g_B^* \, \Gamma(B \rightarrow \chi\bar{\chi}) \int \frac{|\vec{p}_B|^2 d|p_B|}{2\pi^2} e^{-E_B/T} \frac{m_B}{E_B} \\
&= \frac{g_B^* m_B}{2\pi^2} \Gamma(B \rightarrow \chi\bar{\chi}) \int_{m_B}^{\infty} dE_B \sqrt{E_B^2 - m_B^2} \, e^{-E_B/T} \\
&\quad \text{using} \quad \frac{d|\vec{p}_B|}{dE_B} = \frac{E_B dE_B}{\sqrt{E_B^2 - m_B^2}} \\
&= \frac{g_B^* m_B^2}{2\pi^2} \Gamma(B \rightarrow \chi\bar{\chi}) \, T \, K_1(m_B/T) ,
\end{aligned}
\tag{3.77}
$$

where $K_1(z)$ is the modified Bessel function of the second kind. For small z it is approximately given by $K_1(z) \approx 1/z$, while for large z it reproduces the Boltzmann factor, $K_1(z) \propto e^{-z}/z + \mathcal{O}(1/z)$. This form suggests that the dark matter density will increase until T becomes small compared to m_B and the source term becomes suppressed by $e^{-m_B/T}$. The source term is independent of n_χ and proportional to the partial decay width. We also expect it to be proportional to the equilibrium number density of B, defined as

$$
n_B^{\text{eq}} = g_B^* \int \frac{d^3 p}{(2\pi)^3} e^{-E_B/T} = \frac{g_B^*}{2\pi^2} \int_{m_B}^\infty dE_B \, E_B \sqrt{E_B^2 - m_B^2}\, e^{-E_B/T}
$$

$$
= \frac{g_B^*}{2\pi^2} m_B^2 \, T \, K_2(m_B/T) ,
\tag{3.78}
$$

in analogy to Eq. (3.77), but with the Bessel function of the first kind K_1. We can use this relation to eliminate the explicit temperature dependence of the source term,

$$
S(B \to \chi\bar{\chi}) = \Gamma(B \to \bar{\chi}\chi) \frac{K_1(m_B/T)}{K_2(m_B/T)} n_B^{\text{eq}} .
\tag{3.79}
$$

To compute the relic density, we introduce the notation of Eq. (3.25), namely $x = m_B/T$ and $Y = n_\chi/T^3$. The Boltzmann equation from Eq. (3.76) now reads

$$
\frac{dY(x)}{dx} = \frac{g_B^*}{2\pi^2} \frac{\Gamma(B \to \chi\bar{\chi})}{H(x_B = 1)} x^3 K_1(x)
$$

$$
\text{with} \quad x^3 K_1(x) \approx \begin{cases} x^3 e^{-x} & x \gg 1 \text{ or } T \ll m_B \\ x^2 & x \ll 1 \text{ or } T \gg m_B. \end{cases}
\tag{3.80}
$$

Because the function $x^3 K_1(x)$ has a distinct maximum at $x \approx 1$, dark matter production is dominated by temperatures $T \approx m_B$. We can integrate the dark matter production over the entire thermal history and find for the final yield $Y(x'_{\text{dec}})$ with the help of the appropriate integral table

$$
Y(x'_{\text{dec}}) \equiv \frac{g_B^*}{2\pi^2} \frac{\Gamma(B \to \chi\bar{\chi})}{H(x_B = 1)} \int_0^\infty x^3 K_1(x)\, dx = \frac{3 g_B^*}{4\pi} \frac{\Gamma(B \to \chi\bar{\chi})}{H(x_B = 1)} .
\tag{3.81}
$$

We can now follow the steps from Eqs. (3.14) and (3.32) and compute the relic density today,

$$
\Omega_\chi h^2 = \frac{h^2}{3 M_{\text{Pl}}^2 H_0^2} \frac{m_\chi}{28} T_0^3 Y(x'_{\text{dec}})
$$

$$
= \frac{h^2}{112\pi} \frac{m_\chi}{M_{\text{Pl}}^2} \frac{T_0^3}{H^2 H(x_B = 1)} \Gamma(B \to \chi\bar{\chi})
$$

$$
\overset{\text{Eq. (1.47)}}{=} \frac{\sqrt{90} h^2}{112\pi^2} \frac{g_B^*}{\sqrt{g_{\text{eff}}}} \frac{m_\chi}{m_B^2} \frac{T_0^3}{H^2 M_{\text{Pl}}} \Gamma(B \to \chi\bar{\chi})
$$

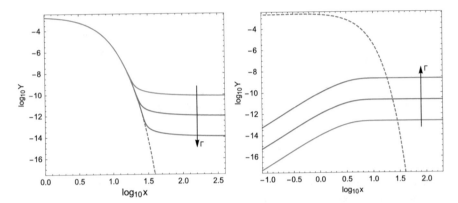

Fig. 3.2 Scaling of $Y(x) = n_\chi/T^3$ for the freeze-out (left) and freeze-in (right) mechanisms for three different interaction rates (larger to smaller cross sections along the arrow). In the left panel $x = m_\chi/T$ and in the right panel $x = m_B/T$. The dashed contours correspond to the equilibrium densities. Figure from Ref. [3]

$$= \frac{\sqrt{90}h^2}{112\pi^2} \frac{g_B^*}{\sqrt{g_{\mathrm{eff}}}} \frac{m_\chi}{m_B^2} \frac{(2.4 \cdot 10^{-4})^3}{(2.5 \cdot 10^{-3})^4} M_{\mathrm{Pl}} \, \Gamma(B \to \chi\bar{\chi})$$

$$= 3.6 \cdot 10^{23} \frac{g_B^* m_\chi}{m_B^2} \Gamma(B \to \chi\bar{\chi}) \,. \tag{3.82}$$

The calculation up to this point is independent from the details of the interaction between the decaying particle B and the DM candidate χ. For the example interaction Eq. (3.75), the partial decay with is given by $\Gamma(B \to \chi\bar{\chi}) = y^2 m_B/(8\pi)$, and assuming $g_B^* = 2$ we find

$$\Omega_\chi h^2 = 0.12 \left(\frac{y}{2 \cdot 10^{-12}}\right)^2 \frac{m_\chi}{m_B} \,. \tag{3.83}$$

The correct relic density from B-decays requires small couplings y and/or dark matter masses m_χ, compatible with the initial assumption that dark matter was never in thermal equilibrium with the Standard Model for $T \gtrsim m_B$. Following Eq. (3.83), larger interaction rates lead to larger final dark matter abundances. This is the opposite scaling as for the freeze-out mechanism of Eq. (3.33). In the right panel of Fig. 3.2 we show the scaling of $Y(x)$ with $x = m_B/T$, compared with the scaling of $Y(x)$ with $x = m_\chi/T$ for freeze-out. Both mechanisms can be understood as the limits of increasing the interaction strength between the visible and the dark matter sector (freeze-out) and decreasing this interaction strength (freeze-in) in a given model.

Even though we illustrate the freeze-in mechanism with the example of the decay of the SM particle B into dark matter, the dark matter sector could also be populated by an annihilation process $B\bar{B} \to \chi\bar{\chi}$, decays of SM particles into a visible particle and dark matter $B \rightarrow B_2\chi$, or scenarios where B is not a SM particle. If the decay $B \rightarrow B_2\chi$ is responsible for the observed relic density, it can account for asymmetric dark matter if $\Gamma(B \rightarrow B_2\chi) \neq \Gamma(\bar{B} \rightarrow \bar{B}_2\bar{\chi})$, as discussed in Sect. 2.5.

References

1. Bellazzini, B., Cliche, M., Tanedo, P.: Effective theory of self-interacting dark matter. Phys. Rev. D **88**(8), 083506 (2013). arXiv:1307.1129 [hep-ph]
2. Lisanti, M.: Lectures on Dark Matter Physics (2016). arXiv:1603.03797 [hep-ph]
3. Bernal, N., Heikinheimo, M., Tenkanen, T., Tuominen, K., Vaskonen, V.: The dawn of FIMP dark matter: a review of models and constraints. Int. J. Mod. Phys. A **32**(27), 1730023 (2017). arXiv:1706.07442 [hep-ph]

Chapter 4
WIMP Models

If we want to approach the problem of dark matter from a particle physics perspective, we need to make assumptions about the quantum numbers of the weakly interacting state which forms dark matter. During most of these lecture notes we assume that this new particle has a mass in the GeV to TeV range, and that its density is thermally produced during the cooling of the Universe. Moreover, we assume that the entire dark matter density of the Universe is due to one stable particle.

The first assumption fixes the spin of this particle. From the Standard Model we know that there exist fundamental scalars, like the Higgs, fundamental fermions, like quarks and leptons, and fundamental gauge bosons, like the gluon or the weak gauge bosons. Scalars have spin zero, fermions have spin 1/2, and gauge bosons have spin 1. Because calculations with gauge bosons are significantly harder, in particular when they are massive, we limit ourselves to scalars and fermions.

When we construct particle models of dark matter we are faced with this wide choice of new, stable particles and their quantum numbers. Moreover, dark matter has to couple to the Standard Model, because it has to annihilate to produce the observed relic density $\Omega_\chi h = 0.12$. This means that strictly speaking we do not only need to postulate a dark matter particle, but also a way for this state to communicate to the Standard Model along the line of the table of states in Sect. 3.5. The second state is usually called a mediator.

4.1 Higgs Portal

An additional scalar particle in the Standard Model can couple to the Higgs sector of the Standard Model in a unique way. The so-called Higgs portal interactions is renormalizable, which means that the coupling constant between two Higgs bosons and two new scalars has a mass unit zero and can be represented by a c-number. All

© Springer Nature Switzerland AG 2019
M. Bauer, T. Plehn, *Yet Another Introduction to Dark Matter*,
Lecture Notes in Physics 959, https://doi.org/10.1007/978-3-030-16234-4_4

we do in such a model is extend the renormalizable Higgs potential of the Standard Model [1], which has a non-zero vacuum expectation value (VEV) for $\mu_H^2 < 0$,

$$V_{\text{SM}} = \mu_H^2 \, \phi^\dagger \phi + \lambda_H (\phi^\dagger \phi)^2 \supset \mu_H^2 \frac{(H + v_H)^2}{2}$$

$$+ \lambda_H \frac{(H + v_H)^4}{4} \supset -\frac{m_H^2}{2} H^2 + \frac{m_H^2}{2v_H} H^3 + \frac{m_H^2}{8v_H^2} H^4 \,, \tag{4.1}$$

In the Standard Model this leads to the two observable mass scales

$$v_H = \sqrt{\frac{-\mu_H^2}{\lambda_H}} = 246 \,\text{GeV} \quad \text{and} \quad m_H = \sqrt{2\lambda_H}\, v_H = 2\sqrt{-\mu_H^2} = 125 \,\text{GeV} \approx \frac{v_H}{2}. \tag{4.2}$$

The last relation is a numerical accident. The general Higgs potential in Eq. (4.1) allows us to couple a new scalar field S to the Standard Model using a renormalizable, dimension-4 term $(\phi^\dagger \phi)(S^\dagger S)$.

For any new scalar field there are two choices we can make. First, we can give it some kind of multiplicative charge, so we actually postulate a set of two particles, one with positive and one with negative charge. This just means that our new scalar field has to be complex values, such that the two charges are linked by complex conjugation. In that case the Higgs portal coupling includes the combination $S^\dagger S$. Alternatively, we can assume that no such charge exists, in which case our new scalar is real and the Higgs portal interaction is proportional to S^2.

Second, we know from the case of the Higgs boson that a scalar can have a finite vacuum expectation value. Due to that VEV, the corresponding new state will mix with the SM Higgs boson to form two mass eigenstates, and modify the SM Higgs couplings and the masses of the W and Z bosons. This is a complication we neither want nor need, so we will work with a dark real scalar. The combined potential reads

$$V = \mu_H^2 \, \phi^\dagger \phi + \lambda_H (\phi^\dagger \phi)^2 + \mu_S^2 \, S^2 + \kappa \, S^3 + \lambda_S \, S^4 + \kappa_3 \phi^\dagger \phi S$$

$$+ \lambda_3 \phi^\dagger \phi S^2 \supset -\frac{m_H^2}{2} H^2 + \frac{m_H^2}{2v_H} H^3 + \frac{m_H^2}{8v_H^2} H^4$$

$$- \mu_S^2 S^2 + \kappa \, S^3 + \lambda_S \, S^4 + \frac{\kappa_3}{2}(H + v_H)^2 S + \frac{\lambda_3}{2}(H + v_H)^2 S^2 \,. \tag{4.3}$$

A possible linear term in the new, real field is removed by a shift in the fields. In the above form the new scalar S can couple to two SM Higgs bosons, which induces a decay either on-shell $S \to HH$ or off-shell $S \to H^* H^* \to 4b$. To forbid this, we apply the usual trick, which is behind essentially all WIMP dark matter models; we require the Lagrangian to obey a global $\mathbb{Z}_2$ symmetry

$$S \to -S, \qquad H \to +H, \qquad \cdots \qquad (4.4)$$

This defines an ad-hoc $\mathbb{Z}_2$ parity $+1$ for all SM particles and -1 for the dark matter candidate. The combined potential now reads

$$
\begin{aligned}
V \supset\; & -\frac{m_H^2}{2} H^2 + \frac{m_H^2}{2v_H} H^3 + \frac{m_H^2}{8v_H^2} H^4 - \mu_S^2 S^2 + \lambda_S S^4 + \frac{\lambda_3}{2}(H + v_H)^2 S^2 \\
=\; & -\frac{m_H^2}{2} H^2 + \frac{m_H^2}{2v_H} H^3 + \frac{m_H^2}{8v_H^2} H^4 \\
& - \left(\mu_S^2 - \lambda_3 \frac{v_H^2}{2}\right) S^2 + \lambda_S S^4 + \lambda_3 v_H\, H S^2 + \frac{\lambda_3}{2} H^2 S^2 \,.
\end{aligned}
\qquad (4.5)
$$

The mass of the dark matter scalar and its phenomenologically relevant SSH and $SSHH$ couplings read

$$
m_S = \sqrt{2\mu_S^2 - \lambda_3 v_H^2} \qquad g_{SSH} = -2\lambda_3 v_H \qquad g_{SSHH} = -2\lambda_3 \,.
\qquad (4.6)
$$

The sign of λ_3 is a free parameter. Unlike for singlet models with a second VEV, the dark singlet does not affect the SM Higgs relations in Eq. (4.2). However, the SSH coupling mediates SS interactions with pairs of SM particles through the light Higgs pole, as well as Higgs decays $H \to SS$, provided the new scalar is light enough. The $SSHH$ coupling can mediate heavy dark matter annihilation into Higgs pairs. We will discuss more details on invisible Higgs decays in Chap. 7.

For dark matter annihilation, the $SSf\bar{f}$ transition matrix element based on the Higgs portal is described by the Feynman diagram

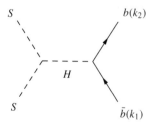

All momenta are defined incoming, giving us for an outgoing fermion and an outgoing anti-fermion

$$
\mathcal{M} = \bar{u}(k_2) \frac{-im_f}{v_H} v(k_1) \frac{-i}{(k_1 + k_2)^2 - m_H^2 + im_H \Gamma_H} (-2i\lambda_3 v_H) \,.
\qquad (4.7)
$$

In this expression we see that v_H cancels, but the fermion mass m_f will appear in the expression for the annihilation rate. We have to square this matrix element, paying attention to the spinors v and u, and then sum over the spins of the external fermions,

$$\sum_{\text{spin}} |\mathcal{M}|^2 = 4\lambda_3^2 m_f^2 \left(\sum_{\text{spin}} v(k_1) \bar{v}(k_1) \right)$$

$$\times \left(\sum_{\text{spin}} u(k_2) \bar{u}(k_2) \right) \frac{1}{\left| (k_1+k_2)^2 - m_H^2 + i m_H \Gamma_H \right|^2}$$

$$= 4\lambda_3^2 m_f^2 \ \text{Tr}\left[(\not{k}_1 - m_f \mathbb{1})(\not{k}_2 + m_f \mathbb{1}) \right] \frac{1}{\left[(k_1+k_2)^2 - m_H^2 \right]^2 + m_H^2 \Gamma_H^2}$$

$$= 4\lambda_3^2 m_f^2 \ 4\left[k_1 k_2 - m_f^2 \right] \frac{1}{\left[(k_1+k_2)^2 - m_H^2 \right]^2 + m_H^2 \Gamma_H^2}$$

$$= 8\lambda_3^2 m_f^2 \ \frac{(k_1+k_2)^2 - 4m_f^2}{\left[(k_1+k_2)^2 - m_H^2 \right]^2 + m_H^2 \Gamma_H^2} \ . \tag{4.8}$$

In the sum over spin and color of the external fermions the averaging is not yet included, because we need to specify which of the external particles are incoming or outgoing. As an example, we compute the cross section for the dark matter annihilation process to a pair of bottom quarks

$$SS \to H^* \to b\bar{b} \ . \tag{4.9}$$

This s-channel annihilation corresponds to the leading on-shell Higgs decay $H \to b\bar{b}$ with a branching ratio around 60%. In terms of the Mandelstam variable $s = (k_1+k_2)^2$ it gives us

$$\overline{\sum_{\text{spin,color}} |\mathcal{M}|^2} = N_c \ 8\lambda_3^2 m_b^2 \ \frac{s - 4m_b^2}{\left(s - m_H^2 \right)^2 + m_H^2 \Gamma_H^2}$$

$$\Rightarrow \quad \sigma(SS \to b\bar{b}) = \frac{1}{16\pi s} \sqrt{\frac{1 - 4m_b^2/s}{1 - 4m_S^2/s}} \ \overline{\sum |\mathcal{M}|^2}$$

$$= \frac{N_c}{2\pi \sqrt{s}} \ \lambda_3^2 m_b^2 \ \sqrt{\frac{1 - 4m_b^2/s}{s - 4m_S^2}} \ \frac{s - 4m_b^2}{\left(s - m_H^2 \right)^2 + m_H^2 \Gamma_H^2} \ . \tag{4.10}$$

To compute the relic density we need the velocity-averaged cross section. For the contribution of the $b\bar{b}$ final state to the dark matter annihilation rate we find the leading term in the non-relativistic limit, $s = 4m_S^2$

$$\langle \sigma v \rangle \Big|_{SS \to b\bar{b}} \equiv \sigma v \Big|_{SS \to b\bar{b}} \overset{\text{Eq. (3.20)}}{=} v \frac{N_c \lambda_3^2 m_b^2}{2\pi \sqrt{s}} \frac{\sqrt{1 - 4m_b^2/s}}{m_S v} \frac{s - 4m_b^2}{\left(s - m_H^2\right)^2 + m_H^2 \Gamma_H^2}$$

$$\overset{\text{threshold}}{=} \frac{N_c \lambda_3^2 m_b^2}{4\pi m_S^2} \sqrt{1 - \frac{m_b^2}{m_S^2}} \frac{4m_S^2 - 4m_b^2}{\left(4m_S^2 - m_H^2\right)^2 + m_H^2 \Gamma_H^2}$$

$$\overset{m_S \gg m_b}{=} \frac{N_c \lambda_3^2 m_b^2}{\pi} \frac{1}{\left(4m_S^2 - m_H^2\right)^2 + m_H^2 \Gamma_H^2} . \tag{4.11}$$

This expression holds for all scalar masses m_S. In our estimate we identify the v-independent expression with the thermal average. Obviously, this will become more complicated once we include the next term in the expansion around $v \approx 0$. The Breit–Wigner propagator guarantees that the rate never diverges, even in the case when the annihilating dark matter hits the Higgs pole in the s-channel.

The simplest parameter point to evaluate this annihilation cross section is on the Higgs pole. This gives us

$$\langle \sigma v \rangle \Big|_{SS \to b\bar{b}} = \frac{N_c \lambda_3^2 m_b^2}{\pi} \frac{1}{\left(4m_S^2 - m_H^2\right)^2 + m_H^2 \Gamma_H^2} \overset{m_H = 2m_S}{=} \frac{N_c \lambda_3^2 m_b^2}{\pi m_H^2 \Gamma_H^2} \approx \frac{15\lambda_3^2}{\text{GeV}^2}$$

$$\Rightarrow \quad \langle \sigma_{\chi\chi} v \rangle = \frac{1}{\text{BR}(H \to b\bar{b})} \langle \sigma v \rangle \Big|_{SS \to b\bar{b}} \approx \frac{25\lambda_3^2}{\text{GeV}^2} \overset{!}{=} 1.7 \cdot 10^{-9} \frac{1}{\text{GeV}^2}$$

$$\Leftrightarrow \quad \boxed{\lambda_3 \approx 8 \cdot 10^{-6}} , \tag{4.12}$$

with $\Gamma_H \approx 4 \cdot 10^{-5} m_H$. While it is correct that the self coupling required on the Higgs pole is very small, the full calculation leads to a slightly larger value $\lambda_3 \approx 10^{-3}$, as shown in Fig. 4.1.

Lighter dark matter scalars also probe the Higgs mediator on-shell. In the Breit-Wigner propagator of the annihilation cross section, Eq. (4.11), we have to compare to the two terms

$$m_H^2 - 4m_S^2 = m_H^2 \left(1 - \frac{4m_S^2}{m_H^2}\right) \quad \Leftrightarrow \quad m_H \Gamma_H \approx 4 \cdot 10^{-5} m_H^2 . \tag{4.13}$$

The two states would have to fulfill exactly the on-shell condition $m_H = 2m_S$ for the second term to dominate. We can therefore stick to the first term for $m_H > 2m_S$

Fig. 4.1 Higgs portal parameter space in terms of the self coupling $\lambda_{hSS} \sim \lambda_3$ and the dark matter mass $M_{DM} = m_S$. The red lines indicate the correct relic density $\Omega_\chi h^2$. Figure from Ref. [2]

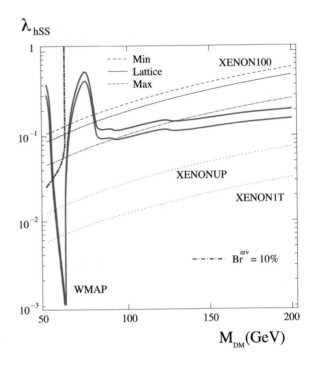

and find for the dominant decay to $b\bar{b}$ pairs in the limit $m_H^2 \gg m_S^2 \gg m_b^2$

$$\langle \sigma v \rangle \Big|_{SS \to b\bar{b}} = \frac{N_c \lambda_3^2 m_b^2}{\pi m_H^4} \approx \frac{\lambda_3^2}{125^2 \, 50^2 \, \text{GeV}^2}$$

$$\overset{!}{=} 1.7 \cdot 10^{-9} \frac{1}{\text{GeV}^2} \quad \Leftrightarrow \quad \boxed{\lambda_3 = 0.26} \, . \tag{4.14}$$

Heavier dark matter scalars well above the Higgs pole also include the annihilation channels

$$SS \to \tau^+ \tau^-, W^+ W^-, ZZ, HH, t\bar{t} \, . \tag{4.15}$$

Unlike for on-shell Higgs decays, the $b\bar{b}$ final state is not dominant for dark matter annihilation when it proceeds through a $2 \to 2$ process. Heavier particles couple to the Higgs more strongly, so above the Higgs pole they will give larger contributions to the dark matter annihilation rate. For top quarks in the final state this simply means replacing the Yukawa coupling m_b^2 by the much larger m_t^2. In addition, the Breit-Wigner propagator will no longer scale like $1/m_H^2$, but proportional to $1/m_S^2$.

Altogether, this gives us a contribution to the annihilation rate of the kind

$$\langle \sigma v \rangle \Big|_{SS \to t\bar{t}} = \frac{N_c \lambda_3^2 m_t^2}{\pi \left(4m_S^2 - m_H^2\right)^2} \overset{2m_S \gg m_H}{=} \frac{N_c \lambda_3^2 m_t^2}{16\pi m_S^4} . \tag{4.16}$$

The real problem is the annihilation to the weak bosons W, Z, because it leads to a different scaling of the annihilation cross section. In the limit of large energies we can describe for example the process $SS \to W^+ W^-$ using spin-0 Nambu-Goldstone bosons in the final state. These Nambu-Goldstone modes in the Higgs doublet ϕ appear as the longitudinal degrees of freedom, which means that dark matter annihilation to weak bosons at large energies follows the same pattern as dark matter annihilation to Higgs pairs. Because we are more used to the Higgs degree of freedom we calculate the annihilation to Higgs pairs,

$$SS \to HH . \tag{4.17}$$

The two Feynman diagrams with the direct four-point interaction and the Higgs propagator at the threshold $s = 4m_S^2$ scale like

$$\mathcal{M}_4 = g_{SSHH} = -2\lambda_3$$

$$\mathcal{M}_H = \frac{g_{SSH}}{s - m_H^2} \frac{3m_H^2}{v_H} \overset{\text{threshold}}{=} -\frac{2\lambda_3 v_H}{4m_S^2 - m_H^2} \frac{3m_H^2}{v_H} \overset{m_S \gg m_H}{=} -\frac{6\lambda_3 m_H^2}{4m_S^2} \ll \mathcal{M}_4 . \tag{4.18}$$

This means for heavy dark matter we can neglect the s-channel Higgs propagator contribution and focus on the four-scalar interaction. In analogy to Eq. (4.11) we then compute the velocity-weighted cross section at threshold,

$$\sigma(SS \to HH) = \frac{1}{16\pi \sqrt{s}} \frac{\sqrt{1 - 4m_H^2/m_S^2}}{\sqrt{s - 4m_S^2}} 4\lambda_3^2 \overset{\text{Eq. (3.20)}}{=} \frac{\lambda_3^2}{4\pi \sqrt{s}} \sqrt{1 - \frac{4m_H^2}{m_S^2}} \frac{1}{v m_S}$$

$$\sigma v \Big|_{SS \to HH} = \frac{\lambda_3^2}{4\pi m_S \sqrt{s}} \sqrt{1 - \frac{4m_H^2}{m_S^2}} \overset{\text{threshold}}{=} \frac{\lambda_3^2}{8\pi m_S^2} \sqrt{1 - \frac{4m_H^2}{m_S^2}} \overset{m_S \gg m_H}{=} \frac{\lambda_3^2}{8\pi m_S^2} \tag{4.19}$$

For $m_S = 200\,\text{GeV}$ we can derive the coupling λ_3 which we need to reproduce the observed relic density,

$$1.7 \cdot 10^{-9} \frac{1}{\text{GeV}^2} \overset{!}{=} \frac{\lambda_3^2}{8\pi m_S^2} \approx \frac{\lambda_3^2}{10^6 \, \text{GeV}^2} \qquad \Leftrightarrow \qquad \boxed{\lambda_3 \approx 0.04} . \tag{4.20}$$

The curve in Fig. 4.1 shows two thresholds related to four-point annihilation channels, one at $m_S = m_Z$ and one at $m_S = m_H$. Starting with $m_S = 200\,\text{GeV}$ and corresponding values for λ_3 the annihilation to Higgs and Goldstone boson pairs dominates the annihilation rate.

One lesson to learn from our Higgs portal considerations is the scaling of the dark matter annihilation cross section with the WIMP mass m_S. It does not follow Eq. (3.3) at all and only follows Eq. (3.35) for very heavy dark matter. For our model, where the annihilation is largely mediated by a Yukawa coupling m_b, we find

$$
\sigma_{\chi\chi} \propto
\begin{cases}
\dfrac{\lambda_3^2 m_b^2}{m_H^4} & m_S \ll \dfrac{m_H}{2} \\[2ex]
\dfrac{\lambda_3^2 m_b^2}{m_H^2 \Gamma_H^2} & m_S = \dfrac{m_H}{2} \\[2ex]
\dfrac{\lambda_3^2}{m_S^2} & m_S > m_Z, m_H \,.
\end{cases}
\tag{4.21}
$$

It will turn out that the most interesting scaling is on the Higgs peak, because the Higgs width is not at all related to the weak scale.

4.2 Vector Portal

Inspired by the WIMP assumption in Eq. (3.3) we can use a new massive gauge boson to mediate thermal freeze-out production. The combination of a free vector mediator mass and a free dark matter mass will allow us to study a similar range scenarios as for the Higgs portal, Eq. (4.21). A physics argument is given by the fact that the Standard Model has a few global symmetries which can be extended to anomaly-free gauge symmetries.

The extension of the Standard Model with its hypercharge symmetry $U(1)_Y$ by an additional $U(1)$ gauge group defines another renormalizable portal to dark matter. Since $U(1)$-field strength tensors are gauge singlets, the kinetic part of the Lagrangian allows for kinetic mixing,

$$
\begin{aligned}
\mathscr{L}_{\text{gauge}} &= -\frac{1}{4}\hat{B}^{\mu\nu}\hat{B}_{\mu\nu} - \frac{s_\chi}{2}\hat{V}^{\mu\nu}\hat{B}_{\mu\nu} - \frac{1}{4}\hat{V}^{\mu\nu}\hat{V}_{\mu\nu} \\[1ex]
&= -\frac{1}{4}\begin{pmatrix}\hat{B}_{\mu\nu} & \hat{V}_{\mu\nu}\end{pmatrix}\begin{pmatrix}1 & s_\chi \\ s_\chi & 1\end{pmatrix}\begin{pmatrix}\hat{B}_{\mu\nu} \\ \hat{V}_{\mu\nu}\end{pmatrix},
\end{aligned}
\tag{4.22}
$$

where $s_\chi \equiv \sin\chi$ is assumed to be a small mixing parameter. In principle it does not have to be an angle, but for the purpose of these lecture notes we assume that it is small, $s_\chi \ll 1$, so we can treat it as a trigonometric function and write $c_\chi \equiv \sqrt{1 - s_\chi}$ and $t_\chi \equiv s_\chi/c_\chi$. Even if the parameter s_χ is chosen to be zero at tree-level, loops of particles charged under both $U(1)_X$ and $U(1)_Y$ introduce a non-zero value for it. Similar to the Higgs portal, there is no symmetry that forbids it, so we do not want to assume that all quantum corrections cancel to a net value $s_\chi = 0$.

The notation $\hat{B}_{\mu\nu}$ indicates that the gauge fields are not yet canonically normalized, which means that the residue of the propagator it not one. In addition, the gauge boson propagators derived from Eq. (4.22) are not diagonal. We can diagonalize the matrix in Eq. (4.22) and keep the hypercharge unchanged with a non-orthogonal rotation of the gauge fields

$$
\begin{pmatrix} \hat{B}_\mu \\ \hat{W}^3_\mu \\ \hat{V}_\mu \end{pmatrix} = G(\theta_V) \begin{pmatrix} B_\mu \\ W^3_\mu \\ V_\mu \end{pmatrix} = \begin{pmatrix} 1 & 0 & -s_\chi/c_\chi \\ 0 & 1 & 0 \\ 0 & 0 & 1/c_\chi \end{pmatrix} \begin{pmatrix} B_\mu \\ W^3_\mu \\ V_\mu \end{pmatrix} . \tag{4.23}
$$

We now include the third component of the $SU(2)_L$ gauge field triplet $W_\mu = (W^1_\mu, W^2_\mu, W^3_\mu)$ which mixes with the hypercharge gauge boson through electroweak symmetry breaking to produce the massive Z boson and the massless photon. Kinetic mixing between the $SU(2)_L$ field strength tensor and the $U(1)_X$ field strength tensor is forbidden because $\hat{V}^{\mu\nu}\hat{A}^a_{\mu\nu}$ is not a gauge singlet. Assuming a mass $\hat{m}_V$ for the V-boson we write the combined mass matrix as

$$
\mathcal{M}^2 \overset{\text{Eq. (4.23)}}{=} \frac{v^2}{4} \begin{pmatrix} g'^2 & -g\,g' & -g'^2 s_\chi \\ -g\,g' & g^2 & g\,g' s_\chi \\ -g'^2 s_\chi & g\,g' s_\chi & \frac{4\hat{m}^2_V}{v^2}(1+s^2_\chi) + g'^2 s^2_\chi \end{pmatrix} + \mathcal{O}(s^3_\chi) . \tag{4.24}
$$

This mass matrix can be diagonalized with a combination of two block-diagonal rotations with the weak mixing matrix and an additional angle ξ,

$$
R_1(\xi) R_2(\theta_w) = \begin{pmatrix} 1 & 0 & 0 \\ 0 & c_\xi & s_\xi \\ 0 & -s_\xi & c_\xi \end{pmatrix} \begin{pmatrix} c_w & s_w & 0 \\ -s_w & c_w & 0 \\ 0 & 0 & 1 \end{pmatrix} , \tag{4.25}
$$

giving

$$
R_1(\xi) R_2(\theta_w) \mathcal{M}^2 R_2(\theta_w)^T R_1(\xi)^T = \begin{pmatrix} m^2_\gamma & 0 & 0 \\ 0 & m^2_Z & 0 \\ 0 & 0 & m^2_V \end{pmatrix} , \tag{4.26}
$$

provided

$$
\tan 2\xi = \frac{2 s_\chi s_w}{1 - \dfrac{\hat{m}^2_V}{\hat{m}^2_Z}} + \mathcal{O}(s^2_\chi) . \tag{4.27}
$$

For this brief discussion we assume for the mass ratio

$$\frac{\hat{m}_V^2}{\hat{m}_Z^2} = \frac{2\hat{m}_V^2}{(g^2 + g'^2)v^2} \ll 1,$$ (4.28)

and find for the physical masses

$$m_\gamma^2 = 0, \qquad m_Z^2 = \hat{m}_Z^2\left[1 + s_\chi^2 s_w^2\left(1 + \frac{\hat{m}_V^2}{\hat{m}_Z^2}\right)\right], \qquad m_V^2 = \hat{m}_V^2\left[1 + s_\chi^2 c_w^2\right].$$ (4.29)

In addition to the dark matter mediator mass we also need the coupling of the new gauge boson V to SM matter. Again, we start with the neutral currents for the not canonically normalized gauge fields and rotate them to the physical gauge bosons defined in Eq. (4.26),

$$\left(ej_{EM}, \frac{e}{\sin\theta_w \cos\theta_w}j_Z, g_D j_D\right)\begin{pmatrix}\hat{A}\\\hat{Z}\\\hat{A}'\end{pmatrix}$$

$$= \left(ej_{EM}, \frac{e}{\sin\theta_w \cos\theta_w}j_Z, g_D j_D\right)K\begin{pmatrix}A\\Z\\V\end{pmatrix},$$ (4.30)

with

$$K = \left[R_1(\xi)R_2(\theta_w)G^{-1}(\theta_\chi)R_2(\theta_w)^{-1}\right]^{-1} \approx \begin{pmatrix}1 & 0 & -s_\chi c_w\\0 & 1 & 0\\0 & s_\chi s_w & 1\end{pmatrix}.$$ (4.31)

The new gauge boson couples to the electromagnetic current with a coupling strength of $-s_\chi c_w e$, while to leading order in s_χ and $\hat{m}_V/\hat{m}_Z$ its coupling to the Z-current vanishes. It is therefore referred to as hidden photon. This behavior changes for larger masses, $\hat{m}_V/\hat{m}_Z \gtrsim 1$, for which the coupling to the Z-current can be the dominating coupling to SM fields. In this case the new gauge boson is called a Z'-boson. For the purpose of these lecture notes we will concentrate on the light V-boson, because it will allow for a light dark matter particle.

There are two ways in which the hidden photon could be relevant from a dark matter perspective. The new gauge boson could be the dark matter itself, or it could provide a portal to a dark matter sector if the dark matter candidate is charged under $U(1)_X$. The former case is problematic, because the hidden photon is not stable and can decay through the kinetic mixing term. Even if it is too light to decay into the

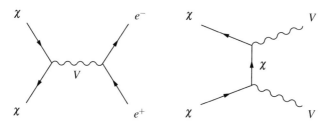

Fig. 4.2 Feynman diagrams contributing to the annihilation of dark matter coupled to the visible sector through a hidden photon

lightest charged particles, electrons, it can decay into neutrinos $V \to \nu\bar{\nu}$ through the suppressed mixing with the Z-boson and into three photons $V \to 3\gamma$ through loops of charged particles. For mixing angles small enough to guarantee stability on time scales of the order of the age of the universe, the hidden photon can therefore not be thermal dark matter.

If the hidden photon couples to a new particle charged under a new $U(1)_X$ gauge group, this particle could be a dark matter candidate. For a new Dirac fermion with $U(1)_X$ charge Q_X, we add a kinetic term with a covariant derivative to the Lagrangian,

$$\mathscr{L}_{\text{DM}} = \bar{\chi} i \gamma^\mu D_\mu \chi - m_\chi \bar{\chi}\chi \qquad \text{with} \quad D_\mu = \partial_\mu - i g_D Q_X V_\mu . \qquad (4.32)$$

Through the $U(1)_X$-mediator this dark fermion is in thermal contact with the Standard Model through the usual annihilation shown in Eq. (3.1). If the dark matter is lighter than the hidden photon and the electron $m_V > m_\chi > m_e$, the dominant s-channel Feynman diagram contributing to the annihilation cross section is shown on the left of Fig. 4.2. This diagram resembles the one shown above Eq. (4.7) for the case of a Higgs portal and the cross section can be computed in analogy to Eq. (4.10),

$$\sigma(\chi\bar{\chi} \to e^+e^-) = \frac{1}{12\pi}(s_\chi c_w e g_D Q_\chi)^2 \left(1 + \frac{2m_e^2}{s}\right)\left(1 + \frac{2m_\chi^2}{s}\right)$$

$$\times \frac{s}{(s - m_V^2)^2 + m_V^2 \Gamma_V^2} \frac{\sqrt{1 - 4\frac{m_e^2}{m_V^2}}}{\sqrt{1 - 4\frac{m_\chi^2}{m_V^2}}} , \qquad (4.33)$$

with Γ_V the total width of the hidden photon V. For the annihilation of two dark matter particles $s = 4m_\chi^2$, and assuming $m_V \gg \Gamma_V$, the thermally averaged annihilation cross section is given by

$$\langle \sigma v \rangle = \frac{1}{4\pi}(s_\chi c_w e g_D Q_\chi)^2 \sqrt{1 - \frac{m_e^2}{m_\chi^2}} \left(1 + \frac{m_e^2}{2m_\chi^2}\right) \frac{4m_\chi^2}{(4m_\chi^2 - m_V^2)^2}$$

$$\overset{m_\chi \gg m_e}{\approx} \frac{m_\chi^2}{\pi m_V^4}(s_\chi c_w e Q_f g_D Q_\chi)^2 . \tag{4.34}$$

It exhibits the same scaling as in the generic WIMP case of Eq. (3.3). In contrast to the WIMP, however, the gauge coupling is rescaled by the mixing angle s_χ and for very small mixing angles the hidden photon can in principle be very light. In Eq. (4.34) we assume that the dark photon decays into electrons. Since the hidden photon branching ratios into SM final states are induced by mixing with the photon, for masses $m_V > 2m_\mu$ the hidden photon also decays into muons and for hidden photon masses above a few 100 MeV and below $2m_\tau$, the hidden photon decays mainly into hadronic states. For $m_V > m_\chi$, the PLANCK bound on the DM mass in Eq. (3.44) implies $m_V > 10$ GeV and Eq. (4.34) would need to be modified by the branching ratios into the different kinematically accessible final states. Instead, we illustrate the scaling with the scenario $Q_\chi = 1$, $m_\chi = 10$ MeV and $m_V = 100$ MeV that is formally excluded by the PLANCK bound, but only allows for hidden photon decays into electrons. In this case, we find the observed relic density given in Eq. (3.32) for a coupling strength

$$\frac{1.7 \cdot 10^{-9}}{\text{GeV}^2} \overset{!}{=} 0.07 \left[\frac{m_\chi}{0.01\,\text{GeV}}\right]^2 \left[\frac{0.1\,\text{GeV}}{m_V}\right]^4 (s_\chi g_D)^2 \quad \Leftrightarrow \quad \boxed{s_\chi g_D = 0.0015} . \tag{4.35}$$

In the opposite case of $m_\chi > m_V > m_e$, the annihilation cross section is dominated by the diagram on the right of Fig. 4.2, with subsequent decays of the hidden photon. The thermally averaged annihilation cross section then reads

$$\langle \sigma v \rangle = \frac{g_D^4 Q_\chi^4}{8\pi} \frac{1}{m_\chi^2} \frac{\left(1 - \frac{m_V^2}{m_\chi^2}\right)^{\frac{3}{2}}}{\left(1 - \frac{m_V^2}{2m_\chi^2}\right)^2} \overset{m_\chi \gg m_V}{\approx} \frac{g_D^4 Q_\chi^4}{8\pi m_\chi^2} . \tag{4.36}$$

The scaling with the dark matter mass is the same as for a WIMP with $m_\chi > m_Z$, as shown in Eq. (3.35). The annihilation cross section is in principle independent of the mixing angle s_χ, motivating the name secluded dark matter for such models, but the hidden photon needs to eventually decay into SM particles. Again assuming

$Q_\chi = 1$, and $m_\chi = 10\,\text{GeV}$, we find

$$\frac{1.7 \cdot 10^{-9}}{\text{GeV}^2} \overset{!}{=} \frac{g_D^4}{8\pi m_\chi^2} = \frac{g_D^4}{250\,\text{GeV}^2} \qquad \Leftrightarrow \qquad \boxed{g_d = 0.025}. \qquad (4.37)$$

4.3 Supersymmetric Neutralinos

Supersymmetry is a (relatively) fashionable model for physics beyond the Standard Model which provides us with a very general set of dark matter candidates. Unlike the portal model described in Sect. 4.1 the lightest supersymmetric partner (LSP) is typically a fermion, more specifically a Majorana fermion. Majorana fermions are their own anti-particles. An on-shell Dirac fermion, like an electron, has four degrees of freedom; for the particle e^- we have two spin directions, and for the anti-particle e^+ we have another two. The Majorana fermion only has two degrees of freedom. The reason why the minimal supersymmetric extension of the Standard Model, the MSSM, limits us to Majorana fermions is that the photon as a massless gauge boson only has two degrees of freedom. This holds for both, the bino partner of the hypercharge gauge boson B and the wino partner of the still massless $SU(2)_L$ gauge boson W^3. Just like the gauge bosons in the Standard Model mix to the photon and the Z, the bino and wino mix to form so-called neutralinos. The masses of the physical state can be computed from the bino mass parameter M_1 and the wino mass parameter M_2.

For reasons which we do not have to discuss in these lecture notes, the MSSM includes a non-minimal Higgs sector: the masses of up-type and down-type fermions are not generated from one Higgs field. Instead, we have two Higgs doublets with two vacuum expectation values v_u and v_d. Because both contribute to the weak gauge boson masses, their squares have to add to

$$v_u^2 + v_d^2 = v_H^2 = (246\,\text{GeV})^2$$

$$\Leftrightarrow \qquad v_u = v_H \cos\beta \qquad v_d = v_H \sin\beta \qquad \Leftrightarrow \qquad \tan\beta = \frac{v_u}{v_d}. \qquad (4.38)$$

Two Higgs doublets include eight degrees of freedom, out of which three Nambu-Goldstone modes are needed to make the weak bosons massive. The five remaining degrees of freedom form a light scalar h^0, a heavy scalar H^0, a pseudo-scalar A^0, and a charged Higgs $H^\pm$. Altogether this gives four neutral and four charged degrees of freedom. In the Standard Model we know that the one neutral (pseudo-scalar) Nambu-Goldstone mode forms one particle with the W^3 gauge bosons. We can therefore expect the supersymmetric higgsinos to mix with the bino and wino as well. Because the neutralinos still are Majorana fermions, the eight degrees of freedom form four neutralino states $\tilde{\chi}_i^0$. Their mass matrix has the form

$$
\mathcal{M} = \begin{pmatrix}
M_1 & 0 & -m_Z s_w \cos\beta & m_Z s_w \sin\beta \\
0 & M_2 & m_Z c_w \cos\beta & -m_Z c_w \sin\beta \\
-m_Z s_w \cos\beta & m_Z c_w \cos\beta & 0 & -\mu \\
m_Z s_w \sin\beta & -m_Z c_w \sin\beta & -\mu & 0
\end{pmatrix}. \tag{4.39}
$$

The mass matrix is real and therefore symmetric. In the upper left corner the bino and wino mass parameters appear, without any mixing terms between them. In the lower right corner we see the two higgsino states. Their mass parameter is μ, the minus sign is conventional; by definition of the Higgs potential it links the up-type and down-type Higgs or higgsino fields, so it has to appear in the off-diagonal entries. The off-diagonal sub-matrices are proportional to m_Z. In the limit $s_w \to 0$ and $\sin\beta = \cos\beta = 1/\sqrt{2}$ a universal mixing mass term $m_Z/\sqrt{2}$ between the wino and each of the two higgsinos appears. It is the supersymmetric counterpart of the combined Goldstone-W^3 mass m_Z.

As any symmetric matrix, the neutralino mass matrix can be diagonalized through a real orthogonal rotation,

$$
N \, \mathcal{M} \, N^{-1} = \mathrm{diag}\left(m_{\tilde{\chi}_j^0}\right) \qquad j = 1, 2 \tag{4.40}
$$

It is possible to extend the MSSM such that the dark matter candidates become Dirac fermions, but we will not explore this avenue in these lecture notes.

Because the $SU(2)_L$ gauge bosons as well as the Higgs doublet include charged states, the neutralinos are accompanied by chargino states. They cannot be Majorana particles, because they carry electric charge. However, as a remainder of the neutralino Majorana property they do not have a well-defined fermion number, like electrons or positrons have. The corresponding chargino mass matrix will not include a bino-like state, so it reads

$$
\mathcal{M} = \begin{pmatrix}
M_2 & \sqrt{2}m_W \sin\beta \\
\sqrt{2}m_W \cos\beta & \mu
\end{pmatrix} \tag{4.41}
$$

It includes the remaining four degrees of freedom from the wino sector and four degrees of freedom from the higgsino sector. As for the neutralinos, the wino and higgsino components mix via a weak mass term. Because the chargino mass matrix is real and not symmetric, it can only be diagonalized using two unitary matrices,

$$
U^* \, \mathcal{M} \, V^{-1} = \mathrm{diag}\left(m_{\tilde{\chi}_j^\pm}\right) \qquad j = 1, 2 \tag{4.42}
$$

For the dark matter phenomenology of the neutralino–chargino sector it will turn out that the mass difference between the lightest neutralino(s) and the lightest chargino are the relevant parameters. The reason is a possible co-annihilation process as described in Sect. 3.3

We can best understand the MSSM dark matter sector in terms of the different $SU(2)_L$ representations. The bino state as the partner of the hypercharge gauge boson is a singlet under $SU(2)_L$. The wino fields with the mass parameter M_2 consist of two neutral degrees of freedom as well as four charged degrees of freedom, one for each polarization of $W^\pm$. Together, the supersymmetric partners of the W boson vector field also form a triplet under $SU(2)_L$. Finally, each of the two higgsinos arise as supersymmetric partner of an $SU(2)_L$ Higgs doublet. The neutralino mass matrix in Eq. (4.39) therefore interpolates between singlet, doublet, and triplet states under $SU(2)_L$.

The most relevant couplings of the neutralinos and charginos we need to consider for our dark matter calculations are

$$g_{Z\tilde\chi_1^0\tilde\chi_1^0} = \frac{g}{2c_w}\left(|N_{13}|^2 - |N_{14}|^2\right)$$

$$g_{h\tilde\chi_1^0\tilde\chi_1^0} = \left(g'N_{11} - gN_{12}\right)(\sin\alpha\, N_{13} + \cos\alpha\, N_{14})$$

$$g_{A\tilde\chi_1^0\tilde\chi_1^0} = \left(g'N_{11} - gN_{12}\right)(\sin\beta\, N_{13} - \cos\beta\, N_{14})$$

$$g_{\gamma\tilde\chi_1^+\tilde\chi_1^-} = e$$

$$g_{W\tilde\chi_1^0\tilde\chi_1^+} = g\left(\frac{1}{\sqrt{2}}N_{14}V_{12}^* - N_{12}V_{11}^*\right), \tag{4.43}$$

with $e = gs_w$, $s_w^2 \approx 1/4$ and hence $c_w^2 \approx 3/4$. The mixing angle α describes the rotation from the up-type and down-type supersymmetric Higgs bosons into mass eigenstates. In the limit of only one light Higgs boson with a mass of 126 GeV it is given by the decoupling condition $\cos(\beta - \alpha) \to 0$. The above form means for those couplings which contribute to the (co-) annihilation of neutralino dark matter

- neutralinos couple to weak gauge bosons through their higgsino content
- neutralinos couple to the light Higgs through gaugino–higgsino mixing
- charginos couple to the photon diagonally, like any other charged particle
- neutralinos and charginos couple to a W-boson diagonally as higgsinos and gauginos

Finally, supersymmetry predicts scalar partners of the quarks and leptons, so-called squarks and sleptons. For the partners of massless fermions, for example squarks $\tilde q$, there exists a $q\tilde q\tilde\chi_j^0$ coupling induced through the gaugino content of the neutralinos. If this kind of coupling should contribute to neutralino dark matter annihilation, the lightest supersymmetric scalar has to be almost mass degenerate with the lightest neutralino. Because squarks are strongly constrained by LHC searches and because of the pattern of renormalization group running, we usually assume one of the sleptons to be this lightest state. In addition, the mixing of the scalar partners of the left-handed and right-handed fermions into mass eigenstates is driven by the corresponding fermion mass, the most attractive co-annihilation scenario in the scalar sector is stau–neutralino co-annihilation. However, in these lecture notes we

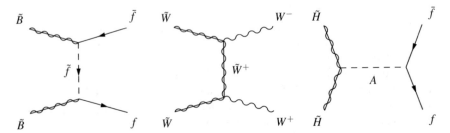

Fig. 4.3 Sample Feynman diagrams for the annihilation of supersymmetric binos (left), winos (center), and higgsinos

will focus on a pure neutralino–chargino dark matter sector and leave the discussion of the squark–quark–neutralino coupling to Chap. 7 on LHC searches.

Similar to the previous section we now compute the neutralino annihilation rate, assuming that in the 10–1000 GeV mass range they are thermally produced. For mostly bino dark matter with

$$M_1 \ll M_2, |\mu| \tag{4.44}$$

the annihilation to the observed relic density is a problem. There simply is no relevant $2 \to 2$ Feynman diagram, unless there is help from supersymmetric scalars $\tilde{f}$, as shown in Fig. 4.3. If for example light staus appear in the t-channel of the annihilation rate we find

$$\sigma(\tilde{B}\tilde{B} \to f\bar{f}) \approx \frac{g^4 m_{\tilde{\chi}_1^0}^2}{16\pi m_{\tilde{f}}^4} \qquad \text{with} \quad m_{\tilde{\chi}_1^0} \approx M_1 \ll m_{\tilde{f}}. \tag{4.45}$$

The problem with pure neutralino annihilation is that in the limit of relatively heavy sfermions the annihilation cross section drops rapidly, leading to a too large predicted bino relic density. Usually, this leads us to rely on stau co-annihilation for a light bino LSP. Along these lines it is useful to mention that with gravity-mediated supersymmetry breaking we assume M_1 and M_2 to be identical at the Planck scale, which given the beta functions of the hypercharge and the weak interaction turns into the condition $M_1 \approx M_2/2$ at the weak scale, i.e. light bino dark matter would be a typical feature in these models.

If M_1 becomes larger than M_2 or μ we can to a good approximation consider the limit

$$M_2 \ll M_1 \to \infty \qquad \text{and} \qquad |\mu| \ll M_1 \to \infty, \tag{4.46}$$

In that case we know that independent of the relation of M_2 and μ there will be at least the lightest chargino close in mass to the LSP. It appears in the t-channel of the actual annihilation diagram and as a co-annihilation partner. To avoid a second neutralino in the co-annihilation process we first consider wino dark matter,

$$M_2 \ll \mu, M_1, m_{\tilde{f}} \;. \tag{4.47}$$

From the list of neutralino couplings in Eq. (4.43) we see that in the absence of additional supersymmetric particles pure wino dark matter can annihilate through a t-channel chargino, as illustrated in Fig. 4.3. Based on the known couplings and on dimensional arguments the annihilation cross section should scale like

$$\sigma(\tilde{W}\tilde{W} \to W^+W^-) \approx \frac{1}{16\pi} \sqrt{\frac{1 - 4m_W^2/s}{1 - 4m_{\tilde{\chi}_1^0}^2/s}} \frac{g^4 s_w^2}{c_w^4 m_{\tilde{\chi}_1^0}^2} \quad \text{with} m_{\tilde{\chi}_1^0} \approx m_{\tilde{\chi}_1^\pm} \approx M_2 \gg m_W$$

$$\approx \frac{1}{16\pi} \frac{1}{m_{\tilde{\chi}_1^0} v} \frac{g^4 s_w^4}{c_w^4 m_{\tilde{\chi}_1^0}^2} \qquad \Rightarrow \qquad \boxed{\sigma_{\chi\chi} \propto \frac{1}{m_{\tilde{\chi}_1^0}^2}} \tag{4.48}$$

The scaling with the mass of the dark matter agent does not follow our original postulate for the WIMP miracle in Eq. (3.3), which was $\sigma_{\chi\chi} \propto m_{\tilde{\chi}_1^0}^2/m_W^4$. If we only rely on the direct dark matter annihilation, the observed relic density translated into a comparably light neutralino mass,

$$\langle \sigma v \rangle \Big|_{\tilde{W}\tilde{W} \to W^+W^-} = \frac{g^4 s_w^4}{16\pi c_w^4 m_{\tilde{\chi}_1^0}^2} \approx \frac{0.7^4}{450\, m_{\tilde{\chi}_1^0}^2} \overset{\text{Eq. (3.35)}}{=} 1.7 \cdot 10^{-9} \frac{1}{\text{GeV}^2}$$

$$\Leftrightarrow \qquad m_{\tilde{\chi}_1^0} \approx 560 \,\text{GeV} \;. \tag{4.49}$$

However, this estimate is numerically poor. The reason is that in contrast to this lightest neutralino, the co-annihilating chargino can annihilate through a photon s-channel diagram into charged Standard Model fermions,

$$\sigma(\tilde{\chi}_1^+ \tilde{\chi}_1^- \to \gamma^* \to f\bar{f}) \approx \sum_f \frac{N_c e^4}{16\pi m_{\tilde{\chi}_1^\pm}^2} = \sum_f \frac{N_c g^4 s_w^2}{16\pi m_{\tilde{\chi}_1^\pm}^2} \;. \tag{4.50}$$

For light quarks alone the color factor combined with the sum over flavors adds a factor $5 \times 3 = 15$ to the annihilation rate. In addition, for $\tilde{\chi}_1^+ \tilde{\chi}_1^-$ annihilation we need to take into account the Sommerfeld enhancement through photon exchange between the slowly moving incoming charginos, as derived in Sect. 3.5. This gives us the correct values

$$\boxed{\Omega_{\tilde{W}} h^2 \approx 0.12 \left(\frac{m_{\tilde{\chi}_1^0}}{2.1\,\text{TeV}}\right)^2 \overset{\text{Sommerfeld}}{\longrightarrow} 0.12 \left(\frac{m_{\tilde{\chi}_1^0}}{2.6\,\text{TeV}}\right)^2} \;. \tag{4.51}$$

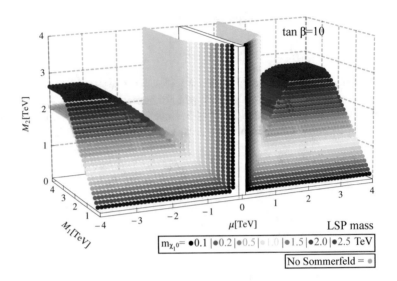

Fig. 4.4 Combinations of neutralino mass parameters M_1, M_2, μ that produce the correct relic abundance, accounting for Sommerfeld-enhancement, along with the LSP mass. The relic surface without Sommerfeld enhancement is shown in gray. Figure from Ref. [3]

In Fig. 4.4 and in the left panel of Fig. 4.5 this wino LSP mass range appears as a horizontal plateau in M_2, with and without the Sommerfeld enhancement. In the right panel of Fig. 4.5 we show the mass difference between the lightest neutralino and the lighter chargino. Typical values for a wino-LSP mass splitting are around $\Delta m = 150\,\text{MeV}$, sensitive to loop corrections to the mass matrices shown in Eqs. (4.39) and (4.41).

Finally, we can study higgsino dark matter in the limit

$$|\mu| \ll M_1, M_2, m_{\tilde{f}} \, . \tag{4.52}$$

Again from Eq. (4.43) we see that in addition to the t-channel chargino exchange, annihilation through s-channel Higgs states is possible. Again, the corresponding Feynman diagrams are shown in Fig. 4.3. At least in the pure higgsino limit with $N_{i3} = N_{i4}$ the two contributions to the $\tilde{H}\tilde{H}Z^0$ coupling cancel, limiting the impact of s-channel Z-mediated annihilation. Still, these channels make the direct annihilation of higgsino dark matter significantly more efficient than for wino dark matter. The Sommerfeld enhancement plays a sub-leading role, because it mostly affects the less relevant chargino co-annihilation,

$$\Omega_{\tilde{H}}h^2 \approx 0.12 \left(\frac{m_{\tilde{\chi}_1^0}}{1.13\,\text{TeV}} \right)^2 \xrightarrow{\text{Sommerfeld}} 0.12 \left(\frac{m_{\tilde{\chi}_1^0}}{1.14\,\text{TeV}} \right)^2 . \tag{4.53}$$

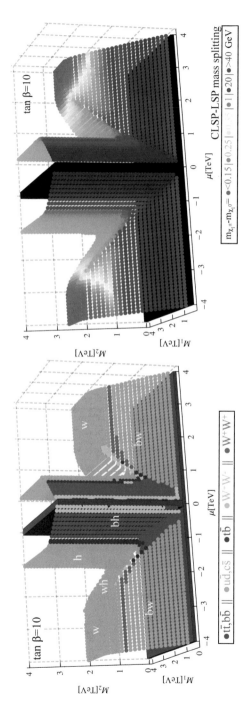

Fig. 4.5 Left: combinations of neutralino mass parameters M_1, M_2, μ that produce the correct relic abundance, not accounting for Sommerfeld-enhancement, along with the leading annihilation product. Parameters excluded by LEP are occluded with a white or black box. Right: mass splitting between the lightest chargino and lightest neutralino. Parameters excluded by LEP are occluded with a white or black box. Figures from Ref. [4]

The higgsino LSP appears in Fig. 4.4 as a vertical plateau in μ. The corresponding mass difference between the lightest neutralino and chargino is much larger than for the wino LSP; it now ranges around a GeV.

Also in Fig. 4.4 we see that a dark matter neutralino in the MSSM can be much lighter than the pure wino and higgsino results in Eqs. (4.51) and (4.53) suggest. For a strongly mixed neutralino the scaling of the annihilation cross section with the neutralino mass changes, and poles in the s-channels appear. In the left panel of Fig. 4.5 we add the leading Standard Model final state of the dark matter annihilation process, corresponding to the distinct parameter regions

- the light Higgs funnel region with $2m_{\tilde{\chi}_1^0} = m_h$. The leading contribution to dark matter annihilation is the decay to b quarks. As a consequence of the tiny Higgs width the neutralino mass has to be finely adjusted. According to Eq. (4.43) the neutralinos couple to the Higgs though gaugino-higgsino mixing. A small, $\mathscr{O}(10\%)$ higgsino component can then give the correct relic density. This very narrow channel with a very light neutralino is not represented in Fig. 4.5. Decays of the Higgs mediator to lighter fermions, like tau leptons, are suppressed by their smaller Yukawa coupling and a color factor;
- the Z-mediated annihilation with $2m_{\tilde{\chi}_1^0} \approx m_Z$, with a final state mostly consisting out of light-flavor jets. The corresponding neutralino coupling requires a sizeable higgsino content. Again, this finely tuned low-mass channel in not shown in Fig. 4.5;
- s-channel annihilation through the higgsino content with some bino admixture also occurs via the heavy Higgs bosons A^0, H^0, and $H^\pm$ with their large widths. This region extends to large neutralino masses, provided the Higgs masses follows the neutralino mass. The main decay channels are $b\bar{b}$, $t\bar{t}$, and $t\bar{b}$. The massive gauge bosons typically decouple from the heavy Higgs sector;
- with a small enough mass splitting between the lightest neutralino and lightest chargino, co-annihilation in the neutralino–chargino sector becomes important. For a higgsino-bino state there appears a large annihilation rate to $\tilde{\chi}_1^0 \tilde{\chi}_1^0 \rightarrow W^+ W^-$ with a t-channel chargino exchange. The wino-bino state will mostly co-annihilate into $\tilde{\chi}_1^0 \tilde{\chi}_1^\pm \rightarrow W^\pm \rightarrow q\bar{q}'$, but also contribute to the $W^+ W^-$ final state. Finally, as shown in Fig. 4.5 the co-annihilation of two charginos can be efficient to reach the observed relic density, leading to a $W^+ W^+$ final state;
- one channel which is absent from our discussion of purely neutralino and chargino dark matter appears for a mass splitting between the scalar partner of the tau lepton, the stau, and the lightest neutralino of few per-cent or less the two states can efficiently co-annihilate. In the scalar quark sector the same mechanism exists for the lightest top squark, but it leads to issues with the predicted light Higgs mass of 126 GeV.

In the right panel of Fig. 4.5 we show the mass difference between the lightest chargino and the lightest neutralino. In all regions where chargino co-annihilation is required, this mass splitting is small. From the form of the mass matrices shown in Eqs. (4.39) and (4.41) this will be the case when either M_2 or μ are the lightest mass parameters. Because of the light higgsino, the two higgsino states in the neutralino sector lead to an additional level separation between the two lightest neutralinos, the degeneracy of the lightest chargino and the lightest neutralino masses will be less precise here. For pure winos the mass difference between the lightest chargino and the lightest neutralino can be small enough that loop corrections matter and the chargino becomes long-lived.

Note that all the above listed channels correspond to ways of enhancing the dark matter annihilation cross section, to allow for light dark matter closer to the Standard Model masses. In that sense they indicate a fine tuning around the generic scaling $\sigma_{\chi\chi} \propto 1/m_{\tilde{\chi}_1^0}^2$ which in the MSSM predicts TeV-scale higgsinos and even heavier winos.

4.4 Effective Field Theory

As another, final theoretical framework to describe the dark matter relic density we introduce an effective theory of dark matter [5]. We will start from the MSSM description and show how the heavy mediator can decouple from the annihilation process. This will put us into a situation similar to the description for example of the muon decay in Fermi's theory. Next, we will generalize this result to an effective Lagrangian. Finally, we will show how this effective Lagrangian describes dark matter annihilation in the early universe.

Let us start with dark matter annihilation mediated by a heavy pseudoscalar A in the MSSM, as illustrated in the right panel of Fig. 4.3. The $A\tilde{\chi}_1^0\tilde{\chi}_1^0$ coupling is defined in Eq. (4.43). If we assume the heavy Higgs to decay to two bottom quarks, the $2 \to 2$ annihilation channel is

$$\tilde{\chi}_1^0 \tilde{\chi}_1^0 \to A^* \to b\bar{b} \,, \qquad (4.54)$$

This description of dark matter annihilation includes two different mass scales, the dark matter mass $m_{\tilde{\chi}_1^0}$ and a decoupled mediator mass $m_A \gg m_{\tilde{\chi}_1^0}$. The matrix element for the dark matter annihilation process includes the A-propagator. From Sect. 3.2 we know that for WIMP annihilation the velocity of the incoming particles is small, $v \ll 1$. If the energy of the scattering process, which determines the momentum flowing through the A-propagator is much smaller than the A-mass, we can approximate the intermediate propagator as

$$\frac{1}{q^2 - m_A^2} \to -\frac{1}{m_A^2} \qquad \Leftrightarrow \qquad \sigma(\tilde{\chi}_1^0 \tilde{\chi}_1^0 \to b\bar{b}) \propto g_{A\tilde{\chi}_1^0\tilde{\chi}_1^0}^2 g_{Abb}^2 \frac{m_b^2}{m_A^4} \,. \qquad (4.55)$$

The fact that the propagator of the heavy scalar A does not include a momentum dependence is equivalent of removing the kinetic term of the A-field from the Lagrangian. We remove the heavy scalar field from the propagating degrees of freedom of our theory. The only actual particles we can use in our description of the annihilation process of Eq. (4.54) are the dark matter fermions $\tilde{\chi}_1^0$ and the bottom quarks. Between them we observe a four-fermion interaction.

On the Lagrangian level, such a four-fermion interactions mediated by a non-propagating state is given by an operator of the type

$$g_{\text{ann}}\, \overline{\psi}_{\tilde{\chi}_1^0} \Gamma^\mu \psi_{\tilde{\chi}_1^0} \overline{\psi}_b \Gamma_\mu \psi_b \,, \qquad (4.56)$$

where $\Gamma^\mu = \{1, \gamma_5, \gamma_\mu, \gamma_\mu\gamma_5, [\gamma_\mu, \gamma_\nu]\}$ represents some kind of Lorentz structure. We know that a Lagrangian has mass dimension four, and a fermion spinor has mass dimension 3/2. The four-fermion interaction then has mass dimension six, and has to be accompanied by a mass-dependent prefactor,

$$\mathscr{L} \supset \frac{g_{\text{ann}}}{\Lambda^2}\, \overline{\psi}_{\tilde{\chi}_1^0} \Gamma^\mu \psi_{\tilde{\chi}_1^0} \overline{\psi}_b \Gamma_\mu \psi_b \,. \qquad (4.57)$$

Given this Lagrangian, the question arises if we want to use this interaction as a simplified description of the MSSM annihilation process or view it as a more general structure without a known ultraviolet completion. For example for the muon decay we nowadays know that the suppression is given by the W-mass of the weak interaction. Using our derivation of Eq. (4.57) we are inspired by the MSSM annihilation channel through a heavy pseudoscalar. In that case the scale Λ should be given by the mass of the lightest particle we integrate out. This defines, modulo order-one factors, the matching condition

$$\Lambda = m_A \qquad \text{and} \qquad g_{\text{ann}} = g_{A\tilde{\chi}_1^0\tilde{\chi}_1^0}\, g_{Abb} \,. \qquad (4.58)$$

From Eq. (4.59) we see that all predictions by the effective Lagrangian are invariant under a simultaneous scaling of the new physics scale Λ and the underlying coupling g_{ann}. Moreover, we know that the annihilation process $\tilde{\chi}_1^0\tilde{\chi}_1^0 \to f\bar{f}$ can be mediated by a scalar in the t-channel. In the limit $m_f \ll m_{\tilde{\chi}_1^0} \ll m_{\tilde{f}}$ this defines essentially the same four-fermion interaction as given in Eq. (4.57).

Indeed, the effective Lagrangian is more general than its interpretation in terms of one half-decoupled model. This suggests to regard the Lagrangian term of Eq. (4.57) as the fundamental description of dark matter, not as an approximation to a full model. For excellent reasons we usually prefer renormalizable Lagrangians, only including operators with mass dimension four or less. Nevertheless, we can extend this approach to examples including all operators up to mass dimension six. This allows to describe all kinds of four-fermion interactions. From constructing the Standard Model Lagrangian we know that given a set of particles we need selection rules to choose which of the possible operators make it into our Lagrangian. Those rules are given by the symmetries of the Lagrangian, local symmetries as well as global symmetries, gauge symmetries as well as accidental symmetries. This way we define a general Lagrangian of the kind

$$\boxed{\mathscr{L} = \mathscr{L}_{\text{SM}} + \sum_j \frac{c_j}{\Lambda^{n-4}} \mathscr{O}_j} ,$$ (4.59)

where the operators $\mathscr{O}_j$ are organized by their dimensionality. The c_j are couplings of the kind shown in Eq. (4.57), called Wilson coefficients, and Λ is the new physics scale.

The one aspect which is crucial for any effective field theory or EFT analysis is the choice of operators contributing to a Lagrangian. Like for any respectable theory we have to assume that any interaction or operator which is not forbidden by a symmetry will be generated, either at tree level or at the quantum level. In practice, this means that any analysis in the EFT framework will have to include a large number of operators. Limits on individual Wilson coefficients have to be derived by marginalizing over all other Wilson coefficients using Bayesian integration (or a frequentist profile likelihood).

From the structure of the Lagrangian we know that there are several ways to generate a higher dimensionality for additional operators,

- external particles with field dimensions adding to more than four. The four-fermion interaction in Eq. (4.57) is one example;
- an energy scale of the Lagrangian normalized to the suppression scale, leading to corrections to lower-dimensional operators of the kind v^2/Λ^2;
- a derivative in the Lagrangian, which after Fourier transformation becomes a four-momentum in the Feynman rule. This gives corrections to lower-dimensional operators of the kind p^2/Λ^2.

For dark matter annihilation we usually rely on dimension-6 operators of the first kind. Another example would be a $\tilde{\chi}_1^0 \tilde{\chi}_1^0 W W$ interaction, which requires a dimension-5 operator if we couple to the gauge boson fields and a dimension-7 operator if we couple to the gauge field strengths. The limitations of an EFT treatment are obvious when we experimentally observe poles, for example the A-resonance in the annihilation process of Eq. (4.54). In the presence of such a resonance it does not help to add higher and higher dimensions—this is similar to Taylor-expanding a pole at a finite energy around zero. Whenever there is a new particle which can be produced on-shell we have to add it to the effective Lagrangian as a new, propagating degree of freedom. Another limiting aspect is most obvious from the third kind of operators: if the correction has the form p^2/Λ^2, and the available energy for the process allows for $p^2 \gtrsim \Lambda^2$, higher-dimensional operators are no longer suppressed. However, this kind of argument has to be worked out for specific observables and models to decide whether an EFT approximation is justified.

Finally, we can estimate what kind of effective theory of dark matter can describe the observed relic density, $\Omega_\chi h^2 \approx 0.12$. As usual, we assume that there is one thermally produced dark matter candidate χ. Two mass scales given by the propagating dark matter agent and by some non-propagating mediator govern our dark matter model. If a dark matter EFT should ever work we need to require that the dark matter mass is significantly smaller than the mediator mass,

$$m_\chi \ll m_{\text{med}} \ . \tag{4.60}$$

In terms of one coupling constant g governing the annihilation process we can use the usual estimate of the WIMP annihilation rate, similar to Eq. (3.3),

$$\langle \sigma_{\chi\chi} \, v \rangle \approx \frac{g^4 m_\chi^2}{4\pi \, m_{\text{med}}^4} \overset{\text{Eq. (3.32)}}{=} \frac{1.7 \cdot 10^{-9}}{\text{GeV}^2} \ . \tag{4.61}$$

We know that it is crucial for this rate to be large enough to bring the thermally produced dark matter rate to the observed level. This gives us a lower limit on the ratio $m_\chi / m_{\text{med}}^2$ or alternatively an upper limit on the mediator mass for fixed dark matter mass. As a rough relation between the mediator and dark matter masses we find

$$\frac{m_{\text{med}}^2}{g^2 m_\chi} = 6.8 \, \text{TeV} \quad \overset{m_\chi = 10 \, \text{GeV}}{\Rightarrow} \quad \frac{m_{\text{med}}}{g} = 260 \, \text{GeV} \gg m_\chi$$

$$\overset{m_\chi = m_{\text{med}}/2}{\Rightarrow} \quad \frac{m_{\text{med}}}{g} = 3.4 \, \text{TeV} = m_\chi \ . \tag{4.62}$$

The dark matter agent in the EFT model can be very light, and the mediator will typically be significantly heavier. An EFT description of dark matter annihilation seems entirely possible.

Going back to our two models, the Higgs portal and the MSSM neutralino, it is less clear if an EFT description of dark matter annihilation works well. In part of the allowed parameter space, dark matter annihilation proceeds through a light Higgs in the s-channel on the pole. Here the mediator is definitely a propagating degree of freedom. For neutralino dark matter we discuss t-channel chargino-mediated annihilation, where $m_{\tilde{\chi}_1^\pm} \approx m_{\tilde{\chi}_1^0}$. Again, the chargino is clearly propagating at the relevant energies.

Finally, to fully rely on a dark matter EFT we need to make sure that all relevant processes are correctly described. For our WIMP models this includes the annihilation predicting the correct relic density, indirect detection and possibly the Fermi galactic center excess introduced in Chap. 5, the limits from direct detection discussed in Chap. 6, and the collider searches of Chap. 7. We will comment on the related challenges in the corresponding sections.

References

1. Plehn, T.: Lectures on LHC Physics. Lect. Notes Phys. 886 (2015). arXiv:0910.4182 [hep-ph]. http://www.thphys.uni-heidelberg.de/~plehn/?visible=review
2. Djouadi, A., Lebedev, O., Mambrini, Y., Quevillon, J.: Implications of LHC searches for Higgs–portal dark matter. Phys. Lett. B **709**, 65 (2014). arXiv:1112.3299 [hep-ph]
3. Bramante, J., Desai, N., Fox, P., Martin, A., Ostdiek, B., Plehn, T.: Towards the final word on neutralino dark matter. Phys. Rev. D **93**(6), 063525 (2016). arXiv:1510.03460 [hep-ph]
4. Bramante, J., Fox, P.J., Martin, A., Ostdiek, B., Plehn, T., Schell, T., Takeuchi, M.: Relic neutralino surface at a 100 TeV collider. Phys. Rev. D **91**, 054015 (2015). arXiv:1412.4789 [hep-ph]
5. Goodman, J., Ibe, M., Rajaraman, A., Shepherd, W., Tait, T.M.P., Yu, H.B.: Constraints on light majorana dark matter from colliders. Phys. Lett. B **695**, 185 (2011). arXiv:1005.1286 [hep-ph]

Chapter 5
Indirect Searches

There exist several ways of searching for dark matter in earth-bound or satellite experiments. All of them rely on the interaction of the dark matter particle with matter, which means they only work if the dark matter particles interacts more than only gravitationally. This is the main assumption of these lecture notes, and it is motivated by the fact that the weak gauge coupling and the weak mass scale happen to predict roughly the correct relic density, as described in Sect. 3.1.

The idea behind indirect searches for WIMPS is that the generally small current dark matter density is significantly enhanced wherever there is a clump of gravitational matter, as for example in the sun or in the center of the galaxy. In these regions dark matter should efficiently annihilate even today, giving us either photons or pairs of particles and anti-particles coming from there. Particles like electrons or protons are not rare, but anti-particles in the appropriate energy range should be detectable. The key ingredient to the calculation of these spectra is the fact that dark matter particles move only very slowly relative to galactic objects. This means we need to compute all processes with incoming dark matter particles essentially at rest. This approximation is even better than at the time of the dark matter freeze-out discussed in Sect. 3.2.

Indirect detection experiments search for many different particles which are produced in dark matter annihilation. First, this might be the particles that dark matter directly annihilated into, for example in a $2 \to 2$ scattering process. This includes protons and anti-protons if dark matter annihilates into quarks. Second, we might see decay products of these particles. An example for such signatures are neutrinos. Examples for dark matter annihilation processes are

$$\tilde{\chi}_1^0 \tilde{\chi}_1^0 \to \ell^+ \ell^-$$

$$\tilde{\chi}_1^0 \tilde{\chi}_1^0 \to q\bar{q} \to p\bar{p} + X$$

$$\tilde{\chi}_1^0 \tilde{\chi}_1^0 \to \tau^+ \tau^-, W^+ W^-, b\bar{b} + X \to \ell^+ \ell^-, p\bar{p} + X \qquad \dots \qquad (5.1)$$

© Springer Nature Switzerland AG 2019
M. Bauer, T. Plehn, *Yet Another Introduction to Dark Matter*,
Lecture Notes in Physics 959, https://doi.org/10.1007/978-3-030-16234-4_5

The final state particles are stable leptons or protons propagating large distances in the Universe. While the leptons or protons can come from many sources, the anti-particles appear much less frequently. One key experimental task in many indirect dark matter searches is therefore the ability to measure the charge of a lepton, typically with the help of a magnetic field. For example, we can study the energy dependence of the antiproton–proton ratio or the positron–electron ratio as a function of the energy. The dark matter signature is either a line or a shoulder in the spectrum, with a cutoff

$$E_{e^+} \approx m_{\tilde{\chi}_1^0} \qquad \text{or} \qquad E_{e^+} < m_{\tilde{\chi}_1^0} \, . \tag{5.2}$$

The main astrophysical background is pulsars, which produce for example electron–positron pairs of a given energy. There exists a standard tool to simulate the propagation of all kinds of particles through the Universe, which is called GAL-PROP. For example Pamela has seen such a shoulder with a positron flux pointing to a very large WIMP annihilation rate. An interpretation in terms of dark matter is inconclusive, because pulsars could provide an alternative explanation and the excess is in tension with PLANCK results from CMB measurements, as discussed in Sect. 3.4.

In these lecture notes we will focus on photons from dark matter annihilation, which we can search for in gamma ray surveys over a wide range of energies. They also will follow one of two kinematic patterns: if they occur in the direct annihilation process, they will appear as a mono-energetic line in the spectrum

$$\chi\chi \to \gamma\gamma \qquad \text{with} \quad E_\gamma \approx m_\chi \, , \tag{5.3}$$

for any weakly interacting dark matter particle χ. This is because the massive dark matter particles are essentially at rest when colliding. If the photons are radiated off charged particles or appear in pion decays $\pi^0 \to \gamma\gamma$

$$\chi\chi \to \tau^+\tau^-, b\bar{b}, W^+W^- \to \gamma + \cdots \, , \tag{5.4}$$

they will follow a fragmentation pattern. We can either compute this photon spectrum or rely on precise measurements from the LEP experiments at CERN (see Sect. 7.1 for a more detailed discussion of the LEP experiments). This photon spectrum will constrain the kind of dark matter annihilation products we should consider, as well as the mass of the dark matter particle.

The energy dependence of the photon flow inside a solid angle $\Delta\Omega$ is given by

$$\frac{d\Phi_\gamma}{dE_\gamma} = \frac{\langle \sigma v \rangle}{8\pi m_{\tilde{\chi}_1^0}^2} \frac{dN_\gamma}{dE_\gamma} \int_{\Delta\Omega} d\Omega \int_l dz \, \rho_\chi^2(z) \, , \tag{5.5}$$

where E_γ is the photon energy, $\langle \sigma v \rangle$ is the usual velocity-averaged annihilation cross-section, N_γ is the number of photons produced per annihilation, and l is the distance from the observer to the actual annihilation event (line of sight). The photon flux depends on the dark matter density squared because it arises from the annihilation of two dark matter particles. A steeper dark matter halo profile, i.e. the dark matter density increasing more rapidly towards the center of the galaxy, results in a more stringent bound on dark matter annihilation. The key problem in the interpretation of indirect search results in terms of dark matter is that we cannot measure the dark matter distributions $\rho_\chi(l)$ for example in our galaxy directly. Instead, we have to rely on numerical simulations of the dark matter profile, which introduce a sizeable parametric or theory uncertainty in any dark-matter related result. Note that the dark matter profile is not some kind of multi-parameter input which we have the freedom to assume freely. It is a prediction of numerical dark matter simulations with associated error bars. Not all papers account for this uncertainty properly. In contrast, the constraints derived from CMB anisotropies discussed in Sect. 1.4 are largely free of astrophysical uncertainties.

There exist three standard density profiles; the steep Navarro-Frenk-White (NFW) profile is given by

$$\rho_{NFW}(r) = \frac{\rho_\odot}{\left(\frac{r}{R}\right)^\gamma \left(1 + \frac{r}{R}\right)^{3-\gamma}} \stackrel{\gamma=1}{=} \frac{\rho_\odot}{\frac{r}{R}\left(1 + \frac{r}{R}\right)^2} , \tag{5.6}$$

where r is the distance from the galactic center. Typical parameters are a characteristic scale $R = 20\,\text{kpc}$ and a solar position dark matter density $\rho_\odot = 0.4\,\text{GeV}/\text{cm}^3$ at $r_\odot = 8.5\,\text{kpc}$. In this form we can easily read off the scaling of the dark matter density in the center of the galaxy, i.e. $r \ll R$; there we find $\rho_{NFW} \propto r^{-\gamma}$. The second steepest is the exponential Einasto profile ,

$$\rho_{Einasto}(r) = \rho_\odot \exp\left[-\frac{2}{\alpha}\left(\left(\frac{r}{R}\right)^\alpha - 1\right)\right] , \tag{5.7}$$

with $\alpha = 0.17$ and $R = 20\,\text{kpc}$. It fits micro-lensing and star velocity data best. Third is the Burkert profile with a constant density inside a radius R,

$$\rho_{Burkert}(r) = \frac{\rho_\odot}{\left(1 + \frac{r}{R}\right)\left(1 + \frac{r^2}{R^2}\right)} , \tag{5.8}$$

where we assume $R = 3\,\text{kpc}$. Assuming a large core results in very diffuse dark matter at the galactic center, and therefore yields the weakest bound on neutralino self annihilation. Instead assuming $R = 0.1\,\text{kpc}$ only alters the dark matter annihilation constraints by an order-one factor. We show the three profiles in Fig. 5.1 and observe that the difference between the Einasto and the NFW parametrizations are marginal, while the Burkert profile has a very strongly reduced dark matter density in the center of the galaxy. One sobering result of this comparison is that whatever theoretical considerations lie behind the NFW and Einasto profiles, once

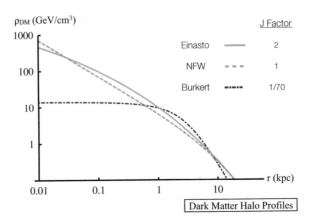

Fig. 5.1 Dark matter galactic halo profiles, including standard Einasto and NFW profiles along with a Burkert profile with a 3 kpc core. J factors are obtained assuming a spherical dark matter distribution and integrating over the radius from the galactic center from $r \simeq 0.05$ to 0.15 kpc. J factors are normalized so that $J(\rho_{NFW}) = 1$. Figure from Ref. [1]

their parameters are fit to data the possibly different underlying arguments play hardly any role. The impact on gamma ray flux of different dark matter halo profiles is conveniently parameterized by the factor

$$ J \propto \int_{\Delta\Omega} d\Omega \int_{\text{line of sight}} dz \, \rho_\chi^2(z) \quad \text{with} \quad J(\rho_{NFW}) \equiv 1 \, . \tag{5.9} $$

Also in Fig. 5.1 we quote the J factors integrated over the approximate HESS galactic center gamma ray search range, $r = 0.05 \ldots 0.15$ kpc. As expected, the Burkert profile predicts a photon flow lower by almost two orders of magnitude. In a quantitative analysis of dark matter signals this difference should be included as a theory error or a parametric error, similar to for example parton densities or the strong coupling in LHC searches.

While at any given time there is usually a sizeable set of experimental anomalies discussed in the literature, we will focus on one of them: the photon excess in the center of our galaxy, observed by Fermi, but discovered in their data by several non-Fermi groups. The excess is shown in Fig. 5.2 and covers the wide photon energy range

$$ E_\gamma = 0.3 \ldots 5 \, \text{GeV} \, , \tag{5.10} $$

and clearly does not form a line. The error bars refer to the interstellar emission model, statistics, photon fragmentation, and instrumental systematics. Note that the statistical uncertainties are dominated not by the number of signal events, but by the statistical uncertainty of the subtracted background events. The fact that uncertainties on photon fragmentation, means photon radiation off other Standard

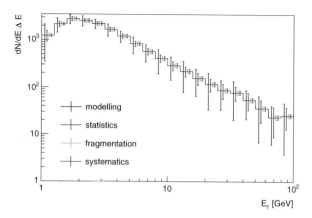

Fig. 5.2 Excess photon spectrum of the Fermi galactic center excess. Figure from Ref. [2], including the original data and error estimates from Ref. [3]

Model particles are included in the analysis, indicates, that for an explanation we resort to photon radiation off dark matter annihilation products, Eq. (5.4). This allows us to link the observed photon spectrum to dark matter annihilation, where the photon radiation off the final state particles is known very well from many collider studies. Two aspects of Fig. 5.2 have to be matched by any explanation. First, the total photon rate has to correspond to the dark matter annihilation rate. It turns out that the velocity-averaged annihilation rate has to be in the same range as the rate required for the observed relic density,

$$\langle \sigma_{\chi\chi} v \rangle = \frac{10^{-8} \dots 10^{-9}}{\text{GeV}^2} \,, \tag{5.11}$$

but with a much lower velocity spectrum now. Second, the energy spectrum of the photons reflects the mass of the dark matter annihilation products. Photons radiated off heavier, non-relativistic states will typically have higher energies. This information is used to derive the preferred annihilation channels given in Fig. 5.3. The official Fermi data confirms these ranges, but with typically larger error bars. As an example, we quote the fit information under the assumption of two dark matter Majorana fermions decaying into a pair of Standard Model states [5]:

Channel	$\langle \sigma_{\chi\chi} v \rangle$ [fb]	m_χ [GeV]
$q\bar{q}$	275 ± 45	24 ± 3
$b\bar{b}$	580 ± 85	49 ± 6
$\tau^+\tau^-$	110 ± 17	10 ± 1
W^+W^-	1172 ± 160	80 ± 1

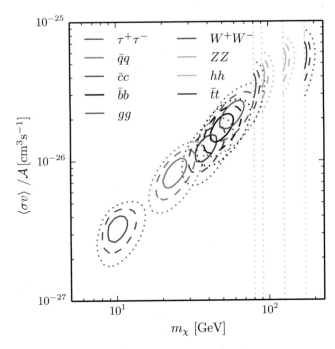

Fig. 5.3 Preferred dark matter masses and cross sections for different annihilation channels [4].
Figure from Ref. [5]

For each of these annihilation channels the question arises if we can also generate
a sizeable dark matter annihilation rate at the center of the galaxy today, while also
predicting the correct relic density $\Omega_\chi h^2$.

5.1 Higgs Portal

Similar to our calculation of the relic density, we will first show what range of
annihilation cross sections from the galactic center can be explained by Higgs portal
dark matter. Because the Fermi data prefers a light dark matter particle we will
focus on the two velocity-weighted cross sections accounting for the observed relic
density and for the galactic center excess around the Higgs pole $m_S/2 = m_H$. First,
we determine how large an annihilation cross section in the galactic center we can
achieve. The typical cross sections given in Eq. (5.11) can be explained by $m_S =
220\,\text{GeV}$ and $\lambda_3 = 1/10$ as well as a more finely tuned $m_S = m_H/2 = 63\,\text{GeV}$
with $\lambda_3 \approx 10^{-3}$, as shown in Fig. 4.1.

We can for example assume that the Fermi excess is due to on-shell Higgs-
mediated annihilation, while the observed relic density does not probe the Higgs
pole. The reason we can separate these two annihilation signals based on the same

Feynman diagram this way is that the Higgs width is smaller than the typical velocities, $\Gamma_H/m_H \ll v$. We start with the general annihilation rate of a dark matter scalar, Eq. (4.10) and express it including the leading relative velocity dependence from Eq. (3.20),

$$s = 4m_S^2 + m_S^2 v^2 = 4m_S^2 \left(1 + \frac{v^2}{4}\right). \tag{5.12}$$

The WIMP velocity at the point of dark matter decoupling in the early universe we find roughly

$$x_{\text{dec}} := \frac{m_S}{T_{\text{dec}}} \overset{\text{Eq. (3.9)}}{=} 28 \quad \Leftrightarrow \quad T_{\text{dec}} \approx \frac{m_S}{28} = \frac{m_S}{2} v_{\text{ann}}^2 \quad \Leftrightarrow \quad v_{\text{ann}}^2 = \frac{1}{14}. \tag{5.13}$$

Today the Universe is colder, and the WIMP velocity is strongly red-shifted. Typical galactic velocities today are

$$v_0 \approx 2.3 \cdot 10^5 \, \frac{\text{m}}{\text{s}} \, \frac{1}{c} \approx \frac{1}{1300} \ll v_{\text{ann}}, \tag{5.14}$$

This hierarchy in typical velocities between the era of thermal dark matter production and annihilation and dark matter annihilation today is what will drive our arguments below.

Only assuming $m_b \ll s$ the general form of the scalar dark matter annihilation rate is

$$
\begin{aligned}
\sigma v \Big|_{SS \to b\bar{b}} &= \frac{N_c}{2\pi} \lambda_3^2 m_b^2 \frac{1}{m_S \sqrt{s}} \frac{s}{(s - m_H^2)^2 + m_H^2 \Gamma_H^2} \\
&= \left(1 + \frac{v^2}{8} + \mathcal{O}(v^4)\right) \frac{N_c}{2\pi} \lambda_3^2 m_b^2 \frac{1}{2m_S^2} \frac{4m_S^2}{(4m_S^2 - m_H^2 + m_S^2 v^2)^2 + m_H^2 \Gamma_H^2} \\
&= \left(1 + \frac{v^2}{8}\right) \frac{N_c}{2\pi} \lambda_3^2 m_b^2 \frac{1}{2m_S^2} \frac{4m_S^2}{(4m_S^2 - m_H^2)^2 + 2(4m_S^2 - m_H^2)m_S^2 v^2 + m_H^2 \Gamma_H^2} \\
&\quad + \mathcal{O}(v^4).
\end{aligned} \tag{5.15}
$$

The typical velocity of the dark matter states only gives a small correction for scalar, s-wave annihilation. It includes two aspects: first, an over-all reduction of the annihilation cross section for finite velocity $v > 0$, and second a combined cutoff of the Breit-Wigner propagator,

$$\max\left[2(4m_S^2 - m_H^2)m_S^2 v^2, m_H^2 \Gamma_H^2\right] = m_S^4 \max\left[8v^2 \left(1 - \frac{m_H^2}{4m_S^2}\right), 16 \cdot 10^{-10}\right]. \tag{5.16}$$

Close to but not on the on-shell pole $m_H = m_S/2$ the modification of the Breit-Wigner propagator can be large even for small velocities, while the rate reduction can clearly not account for a large boost factor describing the galactic center excess. We therefore ignore the correction factor $(1 + v^2/8)$ when averaging the velocity-weighted cross section over the velocity spectrum. If, for no good reason, we assume a narrow Gaussian velocity distribution centered around $\bar{v}$ we can approximate Eq. (5.15) as [6]

$$\langle \sigma v \rangle \Big|_{SS \to b\bar{b}} \approx \frac{N_c}{2\pi} \lambda_3^2 m_b^2 \frac{1}{2m_S^2} \frac{4m_S^2}{\left(4m_S^2 - m_H^2 + \xi\, m_S^2 \bar{v}^2\right)^2 + 4m_S^2 \Gamma_H^2}, \qquad (5.17)$$

with a fitted $\xi \approx 2\sqrt{2}$. This modified on-shell pole condition shifts the required dark matter mass slightly below the Higgs mass $2m_S \lesssim m_H$. The size of this shift depends on the slowly dropping velocity, first at the time of dark matter decoupling, $\bar{v} \equiv v_{\text{ann}}$, and then today, $\bar{v} \equiv v_0 \ll v_{\text{ann}}$. This means that during the evolution of the Universe the Breit-Wigner propagator in Eq. (5.17) is always probed above its pole, probing the actual pole only today.

We first compute the Breit–Wigner suppression of $\langle v\sigma \rangle$ in the early universe, starting with today's on-shell condition responsible for the galactic center excess,

$$m_S \stackrel{!}{=} \frac{m_H}{2\sqrt{1 + \dfrac{v_0^2}{\sqrt{2}}}} \approx \frac{m_H}{2} \quad \Rightarrow \quad 4m_S^2 - m_H^2 + \xi\, m_S^2 v_{\text{ann}}^2$$

$$= 4m_S^2 \left(1 + \frac{v_{\text{ann}}^2}{\sqrt{2}}\right) - m_H^2$$

$$= 4m_S^2 \left(\frac{v_{\text{ann}}^2}{\sqrt{2}} - \frac{v_0^2}{\sqrt{2}}\right)$$

$$\stackrel{v_{\text{ann}} \gg v_0}{\approx} \frac{m_S^2}{5}. \qquad (5.18)$$

This means that the dark matter particle has a mass just slightly below the Higgs pole. Using Eq. (5.17) the ratio of the two annihilation rates, for all other parameters constant, then becomes

$$\frac{\langle \sigma_0 v \rangle}{\langle \sigma_{\text{ann}} v \rangle} = \frac{8m_S^4 v_{\text{ann}}^4}{4m_S^2 \Gamma_H^2} = \frac{2v_{\text{ann}}^4}{16 \cdot 10^{-10}} = \frac{1}{8}\frac{1}{14^2} 10^{10} \gtrsim 10^6. \qquad (5.19)$$

This is the maximum enhancement we can generate to explain Fermi's galactic center excess. The corresponding Higgs coupling λ_3 is given in Fig. 4.1.

We can turn the question around and compute the smallest annihilation cross section in the galactic center consistent with the observed relic abundance in the

Higgs portal model. For this purpose we assume that unlike in Eq. (5.18) the pole condition is fulfilled in the early universe, leading to a Breit-Wigner suppression today of

$$m_S \overset{!}{=} \frac{m_H}{2\sqrt{1 + \frac{v_{\text{ann}}^2}{\sqrt{2}}}}$$

$$\Rightarrow \quad 4m_S^2 - m_H^2 + \xi\, m_S^2 v_0^2 = 4m_S^2 \left(1 + \frac{v_0^2}{\sqrt{2}}\right) - m_H^2 \overset{v_{\text{ann}} \gg v_0}{\approx} -\frac{m_S^2}{5} .$$

$$(5.20)$$

This gives us a ratio of the two velocity-mediated annihilation rates

$$\frac{\langle \sigma_0 v \rangle}{\langle \sigma_{\text{ann}} v \rangle} = \frac{4m_S^2 \Gamma_H^2}{8m_S^4 v_{\text{ann}}^4} \overset{\text{Eq. (5.19)}}{\lesssim} 10^{-6}$$

$$\text{for} \quad m_S \approx \frac{m_H}{2}\left(1 - \frac{v_{\text{ann}}^2}{2\sqrt{2}}\right) \overset{\text{Eq. (5.13)}}{=} 62.91\,\text{GeV} . \qquad (5.21)$$

The dark matter particle now has a mass further below the pole. This means that we can interpolate between the two extreme ratios of velocity-averaged annihilation rates using a very small range of $m_S < m_H/2$. If we are willing to tune this mass relation we can accommodate essentially any dark matter annihilation rate today with the Higgs portal model, close to on-shell Higgs pole annihilation. The key to this result is that following Eq. (5.16) the Higgs width-to-mass ratio is small compared to v_{ann}, so we can decide to assign the on-shell condition to each of the two relevant annihilation processes. In between, none of the two processes will proceed through the on-shell Higgs propagator, which indeed gives $\langle \sigma_0 v \rangle \approx \langle \sigma_{\text{ann}} v \rangle$. The corresponding coupling λ_3 we can read off Fig. 4.1. Through this argument it becomes clear that the success of the Higgs portal model rests on the wide choice of scalings of the dark matter annihilation rate, as shown in Eq. (4.21).

5.2 Supersymmetric Neutralinos

An explanation of the galactic center excess has to be based on the neutralino mass matrix given in Eq. (4.39), defining a dark matter Majorana fermion as a mixture of the bino singlet, the wino triplet, and two higgsino doublets. Some of its relevant couplings are given in Eq. (4.43). Correspondingly, some annihilation processes leading to the observed relic density and underlying our interpretation of the Fermi galactic center excess are illustrated in Fig. 4.3. One practical advantage of the MSSM is that it offers many neutralino parameter regions to play with. We know that pure wino or higgsino dark matter particles reproducing the observed relic density are much heavier than the Fermi data suggests. Instead of these pure states we will rely on mixed states. A major obstacle of all MSSM interpretations are the mass ranges shown in Fig. 5.3, indicating a clear preference of the galactic center excess for neutralino masses $m_{\tilde{\chi}_1^0} \lesssim 60$ GeV. This does not correspond to the typical MSSM parameter ranges giving us the correct relic density. This means that in an MSSM analysis of the galactic center excess the proper error estimate for the photon spectrum is essential.

We start our discussion with the finely tuned annihilation through a SM-like light Higgs or through a Z-boson, i.e. $\tilde{\chi}_1^0 \tilde{\chi}_1^0 \to h^*, Z^* \to b\bar{b}$. The properties of this channel are very similar to those of the Higgs portal. On the left y-axes of Fig. 5.4 we show the (inverse) relic density for a bino-higgsino LSP, both for a wide range of neutralino masses and zoomed into the Higgs pole region. We decouple the wino to $M_2 = 700$ GeV and vary M_1 to give the correct relic density for three fixed, small higgsino mass values. We see that the $b\bar{b}$ annihilation channel only predicts the correct relic density in the two pole regions of the MSSM parameter space, with $m_{\tilde{\chi}_1^0} = 46$ GeV and $m_{\tilde{\chi}_1^0} = 63$ GeV. The width of both peaks is given by the momentum smearing through velocity spectrum rather than physical Higgs width and Z-width. The enhancement of the two peaks over the continuum is comparable, with the Z-funnel coupled to the velocity-suppressed axial-vector current and the Higgs funnel suppressed by the small bottom Yukawa coupling.

On the right y-axis of Fig. 5.4, accompanied by dashed curves, we show the annihilation rate in the galactic center. The rough range needed to explain the Fermi excess is indicated by the horizontal line. As discussed for the Higgs portal, the difference to the relic density is that the velocities are much smaller, so the widths of the peaks are now given by the physical widths of the two mediators. The scalar Higgs resonance now leads to a much higher peak than the velocity-suppressed axial-vector coupling to the Z-mediator. This implies that continuum annihilation as well as Z-pole annihilation would not explain the galactic center excess, while the Higgs pole region could.

Fig. 5.4 Inverse relic density (solid, left axis) and annihilation rate in the galactic center (dashed, right axis) for an MSSM parameter point where the annihilation is dominated by $\tilde{\chi}_1^0 \tilde{\chi}_1^0 \to b\bar{b}$. Figure from Ref. [2]

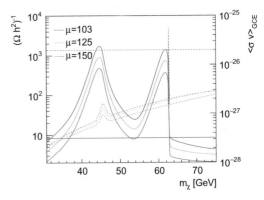

This is why in the right panel of Fig. 5.4 we zoom into the Higgs peak regime. A valid explanation of the galactic center excess requires the solid relic density curves to cross the solid horizontal line and at the same time the dashed galactic center excess lines to cross the dashed horizontal line. We see that there exist finely tuned regions around the Higgs pole which allow for an explanation of the galactic center excess via a thermal relic through the process $\tilde{\chi}_1^0 \tilde{\chi}_1^0 \to b\bar{b}$. The physics of this channel is very similar to scalar Higgs portal dark matter.

For slightly larger neutralino masses, the dominant annihilation becomes $\tilde{\chi}_1^0 \tilde{\chi}_1^0 \to WW$, mediated by a light t-channel chargino combined with chargino-neutralino co-annihilation for the relic density. Equation (4.43) indicates that in this parameter region the lightest neutralino requires either a wino content or a higgsino content. In the left panel of Fig. 5.5 we show the bino-higgsino mass plane indicating the preferred regions from the galactic center excess. The lightest neutralino mass varies from $m_{\tilde{\chi}_1^0} \approx 50\,\text{GeV}$ to more than 250 GeV. Again, we decouple the wino to $M_2 = 700\,\text{GeV}$, so the LSP is a mixture of higgsino, coupling to electroweak bosons, and bino. For this slice in parameter space an increase in $|\mu|$ compensates any increase in M_1, balancing the bino and higgsino contents. The MSSM parameter regions which allow for efficient dark matter annihilation into gauge bosons are strongly correlated in M_1 and μ, but not as tuned as the light Higgs funnel region with its underlying pole condition. Around $M_1 = |\mu| = 200\,\text{GeV}$ a change in shape occurs. It is caused by the on-set of neutralino annihilation to top pairs, in spite of a heavy Higgs mass scale of 1 TeV.

To trigger a large annihilation rate for $\tilde{\chi}_1^0 \tilde{\chi}_1^0 \to t\bar{t}$ we lower the heavy pseudoscalar Higgs mass to $m_A = 500\,\text{GeV}$. In the right panel of Fig. 5.5 we show the preferred parameter range again in the bino-higgsino mass plane and for heavy winos, $M_2 = 700\,\text{GeV}$. As expected, for $m_{\tilde{\chi}_1^0} > 175\,\text{GeV}$ the annihilation into top

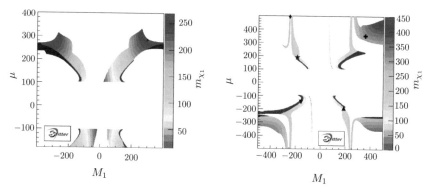

Fig. 5.5 Left: lightest neutralino mass based on the Fermi photon where $\tilde{\chi}_1^0 \tilde{\chi}_1^0 \to WW$ is a dominant annihilation channel. Right: lightest neutralino mass based on the Fermi photon spectrum for $m_A = 500\,\text{GeV}$, where we also observe $\tilde{\chi}_1^0 \tilde{\chi}_1^0 \to t\bar{t}$. The five symbols indicate local best-fitting parameter points. The black shaded regions are excluded by the Fermi limits from dwarf spheroidal galaxies

pairs follows the WW annihilation region in the mass plane. The main difference between the WW and $t\bar{t}$ channels is the smaller M_1 values around $|\mu| = 200\,\text{GeV}$. The reason is that an increased bino fraction compensates for the much larger top Yukawa coupling. The allowed LSP mass range extends to $m_{\tilde{\chi}_1^0} \gtrsim 200\,\text{GeV}$.

The only distinctive feature for $m_A = 500\,\text{GeV}$ in the M_1 vs μ plane is the set of peaks around $M_1 \approx 300\,\text{GeV}$. Here the lightest neutralino mass is around $250\,\text{GeV}$, just missing the A-pole condition. Because on the pole dark matter annihilation through a $2 \rightarrow 1$ process becomes too efficient, the underlying coupling is reduced by a smaller higgsino fraction of the LSP. The large-$|M_1|$ regime does not appear in the upper left corner of Fig. 5.5 because at tree level this parameter region features $m_{\tilde{\chi}_1^+} < m_{\tilde{\chi}_1^0}$ and we have to include loop corrections to revive it.

In principle, for $m_{\tilde{\chi}_1^0} > 126\,\text{GeV}$ we should also observe neutralino annihilation into a pair of SM-like Higgs bosons. However, the t-channel neutralino diagram which describes this process will typically be overwhelmed by the annihilation to weak bosons with the same t-channel mediator, shown in Fig. 4.3. From the annihilation into top pairs we know that s-channel mediators with $m_{A,H} \approx 2m_h$ are in principle available, and depending on the MSSM parameter point the heavy scalar Higgs can have a sizeable branching ratio into two SM-like Higgses. For comparably large velocities in the early universe both s-channel mediators indeed work fine to predict the observed relic density. For the smaller velocities associated with the galactic center excess the CP-odd mediator A completely dominates, while the CP-even H is strongly velocity-suppressed. On the other hand, only the latter couples to two light Higgs bosons, so an annihilation into Higgs pairs responsible for the galactic center excess is difficult to realize in the MSSM.

Altogether we see that the annihilation channels

$$\tilde{\chi}_1^0 \tilde{\chi}_1^0 \rightarrow b\bar{b}, WW, t\bar{t} \qquad \text{with} \quad m_{\tilde{\chi}_1^0} = 63 \ldots 250\,\text{GeV} \tag{5.22}$$

can explain the Fermi galactic center excess and the observed relic density in the MSSM. Because none of them correspond to the central values of a combined fit to the galactic center excess, it is crucial that we take into account all sources of (sizeable) uncertainties. An additional issue which we will only come to in Chap. 6 is that direct detection constraints in addition to requiring the correct relic density and the correct galactic center annihilation rate is a serious challenge to the MSSM explanations.

5.3 Next-to-Minimal Neutralino Sector

An obvious way out of the MSSM limitations is to postulate an addition particle in the s-channel of the neutralino annihilation process and a new, lighter dark matter fermion. This leads us to the next-to-minimal supersymmetric extension, the NMSSM. It introduces an additional singlet under all Standard Model gauge transformations, with its singlino partner. The singlet state forms a second scalar H_1 and a second pseudo-scalar A_1, which will appear in the dark matter annihilation process. As singlets they will only couple to gauge bosons through mixing with the Higgs fields, which guarantees that they are hardly constrained by many searches. The singlino will add a fifth Majorana state to the neutralino mass matrix in Eq. (4.39),

$$
\mathcal{M} = \begin{pmatrix}
M_1 & 0 & -m_Z \cos\beta s_w & m_Z \sin\beta s_w & 0 \\
0 & M_2 & m_Z \cos\beta c_w & -m_Z \sin\beta c_w & 0 \\
-m_Z \cos\beta s_w & m_Z \cos\beta c_w & 0 & -\mu & -m_Z \sin\beta\tilde{\lambda} \\
m_Z \sin\beta s_w & -m_Z \sin\beta c_w & -\mu & 0 & -m_Z \cos\beta\tilde{\lambda} \\
0 & 0 & -m_Z \sin\beta\tilde{\lambda} & -m_Z \cos\beta\tilde{\lambda} & 2\tilde{\kappa}\mu
\end{pmatrix}.
$$

$$(5.23)$$

The singlet/singlino sector can be described by two parameters, the mass parameter $\tilde{\kappa}$ and the coupling for example to the other neutralinos $\tilde{\lambda}$ [7]. First, we need to include the singlino in our description of the neutralino sector. While the wino and the two higgsinos form a triplet or two doublets under $SU(2)_L$, the singlino just adds a second singlet under $SU(2)_L$. The only difference to the bino is that the singlino is also a singlet under $U(1)_Y$, which makes no difference unless we consider co-annihilation driven by hypercharge interaction. A singlet neutralino will therefore interact and annihilate to the observed relic density through its mixing with the wino or with the higgsinos, just like the usual bino.

What is crucial for the explanation of the galactic center excess is the s-channel dark matter annihilation through the new pseudoscalar,

$$
\tilde{\chi}_1^0 \tilde{\chi}_1^0 \to A_1 \to b\bar{b} \quad \text{with} \quad m_{\tilde{\chi}_1^0} = \frac{m_{A_1}}{2} \approx 50 \,\text{GeV}
$$

$$
g_{A_1^0 \tilde{\chi}_1^0 \tilde{\chi}_1^0} = \sqrt{2}\, g\tilde{\lambda} \left(N_{13} N_{14} - \tilde{\kappa} N_{15}^2 \right). \tag{5.24}
$$

We can search for these additional singlet and singlino states at colliders. One interesting aspect is the link between the neutralino and the Higgs sector, which can be probed by looking for anomalous Higgs decays, for example into a pair of dark matter particles. Because an explanation of the galactic center excess requires the singlet and the singlino to be light and to mix with their MSSM counterparts, the resulting invisible branching ratio of the Standard-Model-like Higgs boson can be large.

5.4 Simplified Models and Vector Mediator

The discussion of the dark matter annihilation processes responsible for today's dark matter density as well as a possible galactic center excess nicely illustrates the limitations of the effective theory approach introduced in Sect. 4.4. To achieve the currently observed density with light WIMPs we have to rely on an efficient annihilation mechanism, which can be most clearly seen in the MSSM. For example, we invoke s-channel annihilation or co-annihilation, both of which are not well captured by an effective theory description with a light dark matter state and a heavy, non-propagating mediator. In the effective theory language of Sect. 4.4 this means the mediators are not light compared to the dark matter agent,

$$m_\chi \lesssim m_{\text{med}} . \tag{5.25}$$

In addition, the MSSM and the NMSSM calculations illustrate how one full model extending the Standard Model towards large energy scales can offer several distinct explanations, only loosely linked to each other. In this situation we can collect all the necessary degrees of freedom in our model, but ignore additional states for example predicted by an underlying supersymmetry of the Lagrangian. This approach is called simplified models. It typically describes the dark matter sector, including co-annihilating particles, and a mediator coupling the dark matter sector to the Standard Model. In that language we have come across a sizeable set of simplified models in our explanation of the Fermi galactic center excess:

- dark singlet scalar with SM Higgs mediator (Higgs portal, $SS \to b\bar{b}$);
- dark fermion with SM Z mediator (MSSM, $\tilde{\chi}_1^0 \tilde{\chi}_1^0 \to f\bar{f}$, not good for galactic center excess);
- dark fermion with SM Higgs mediator (MSSM, $\tilde{\chi}_1^0 \tilde{\chi}_1^0 \to b\bar{b}$);
- dark fermion with t-channel fermion mediator (MSSM, $\tilde{\chi}_1^0 \tilde{\chi}_1^0 \to WW$);
- dark fermion with heavy s-channel pseudo-scalar mediator (MSSM, $\tilde{\chi}_1^0 \tilde{\chi}_1^0 \to t\bar{t}$);
- dark fermion with light s-channel pseudo-scalar mediator (NMSSM, $\tilde{\chi}_1^0 \tilde{\chi}_1^0 \to b\bar{b}$).

In addition, we encountered a set of models in our discussion of the relic density in the MSSM in Sect. 4.3:

- dark fermion with fermionic co-annihilation partner and charged s-channel mediator (MSSM, $\tilde{\chi}_1^0 \tilde{\chi}_1^- \to \bar{t}b$);
- dark fermion with fermionic co-annihilation partner and SM W-mediator (MSSM, $\tilde{\chi}_1^0 \tilde{\chi}_1^- \to \bar{u}d$);
- dark fermion with scalar t-channel mediator (MSSM, $\tilde{\chi}_1^0 \tilde{\chi}_1^0 \to \tau\tau$);
- dark fermion with scalar co-annihilation partner (MSSM, $\tilde{\chi}_1^0 \tilde{\tau} \to \tau^*$)

Strictly speaking, all the MSSM scenarios require a Majorana fermion as the dark matter candidate, but we can replace it with a Dirac neutralino in an extended supersymmetric setup.

One mediator which is obviously missing in the above list is a new, heavy vector V or axial-vector. Heavy gauge bosons are ubiquitous in models for physics beyond the Standard Model, and the only question is how we would link or couple them to a dark matter candidate. In principle, there exist different mass regimes in the

$m_\chi - m_V$ mass plane,

$$
\begin{aligned}
m_V > 2m_\chi && \text{possible effective theory} \\
m_V \approx 2m_\chi && \text{on-shell, simplified model} \\
m_V < 2m_\chi && \text{light mediator, simplified model .}
\end{aligned}
\tag{5.26}
$$

To allow for a global analysis including direct detection as well as LHC searches, we couple the vector mediator to a dark matter fermion χ and the light up-quarks,

$$
\mathscr{L} \supset g_u \, \bar{u} \, \gamma^\mu V_\mu \, u + g_\chi \, \bar{\chi} \, \gamma^\mu V_\mu \, \chi \ .
\tag{5.27}
$$

The typical mediator width for $m_\chi \ll m_V$ is

$$
\frac{\Gamma_V}{m_V} \lesssim 0.4 \ \dots \ 10\% \qquad \text{for} \quad g_u = g_\chi = 0.2 \ \dots \ 1 \ .
\tag{5.28}
$$

Based on the annihilation process

$$
\chi\chi \to V^* \to u\bar{u}
\tag{5.29}
$$

we can compute the predicted relic density or the indirect detection prospects. While the $\chi - \chi - V$ interaction also induces a t-channel process $\chi\chi \to V^*V^*$, its contribution to the total dark matter annihilation rate is always strongly suppressed by its 4-body phase space. The on-shell annihilation channel

$$
\chi\chi \to VV
\tag{5.30}
$$

becomes important for $m_V < m_\chi$, with a subsequent decay of the mediator for example to two Standard Model fermions. In that case the dark matter annihilation rate becomes independent of the mediator coupling to the Standard Model, giving much more freedom to avoid experimental constraints.

In Fig. 5.6 we observe that for a light mediator the predicted relic density is smaller than the observed values, implying that the annihilation rate is large. In the left panel we see the three kinematic regimes defined in Eq. (5.26). First, for small mediator masses the $2 \to 2$ annihilation process is $\chi\chi \to u\bar{u}$. The dependence on the light mediator mass is small because the mediator is always off-shell and the position of its pole is far away from the available energy of the incoming dark matter particles. Around the pole condition $2m_\chi \approx m_V \pm \Gamma_V$ the model predicts the correct relic density with very small couplings. For heavy mediators the $2 \to 2$ annihilation process rapidly decouples with large mediator masses, as follows for example from Eq. (3.3). In the right panel of Fig. 5.6 we assume a constant mass ratio $m_V/m_\chi \gtrsim 1$, finding that our simplified vector model has no problems predicting the correct relic density over a wide range of model parameters.

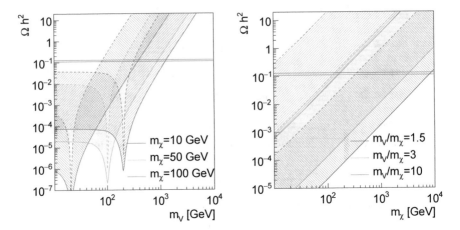

Fig. 5.6 Relic density for the simplified vector mediator model of Eq. (5.27) as a function of the mediator mass for constant dark matter mass (left) and as a function of the dark matter mass for a constant ratio of mediator to dark matter mass (right). Over the shaded bands we vary the couplings $g_u = g_\chi = 0.2, \ldots, 1$. Figure from Ref. [8]

One issue we can illustrate with this non-MSSM simplified model is a strong dependence of our predictions on the assumed model features. The Lagrangian of Eq. (5.27) postulates a coupling to up-quarks, entirely driven by our goal to link dark matter annihilation with direct detection and LHC observables. From a pure annihilation perspective we can also define the mediator coupling to the Standard Model through muons, without changing any of the results shown in Fig. 5.6. Coupling to many SM fermions simultaneously, as we expect from an extra gauge group, will increase the predicted annihilation rate easily by an order of magnitude. Moreover, it is not clear how the new gauge group is related to the $U(1)_Y \times SU(2)_L$ structure of the electroweak Standard Model. All this reflects the fact that unlike the Higgs portal model or supersymmetric extensions a simplified model is hardly more than a single tree-level or loop-level Feynman diagram describing dark matter annihilation. It describes the leading effects for example in dark matter annihilation based on $2 \to 2$ or $2 \to 1$ kinematics or the velocity dependence at threshold. However, because simplified models are usually not defined on the full quantum level, they leave a long list of open questions. For new gauge bosons, also discussed in Sect. 4.2, they include fundamental properties like gauge invariance, unitarity, or freedom from anomalies.

References

1. Bramante, J., Desai, N., Fox, P., Martin, A., Ostdiek, B., Plehn, T.: Towards the final word on neutralino dark matter. Phys. Rev. D **93**(6), 063525 (2016). arXiv:1510.03460 [hep-ph]
2. Butter, A., Murgia, S., Plehn, T., Tait, T.M.P.: Saving the MSSM from the galactic center excess. Phys. Rev. D **96**(3), 035036 (2017). arXiv:1612.07115 [hep-ph]

3. Murgia, S., et al.: [Fermi-LAT Collaboration], Fermi-LAT observations of high-energy γ-ray emission toward the galactic center. Astrophys. J. **819**(1), 44 (2016). arXiv:1511.02938 [astro-ph.HE]
4. Berlin, A., Hooper, D., McDermott, S.D.: Simplified dark matter models for the galactic center gamma-ray excess. Phys. Rev. D **89**(11), 115022 (2014). arXiv:1404.0022 [hep-ph]
5. Calore, F., Cholis, I., McCabe, C., Weniger, C.: A tale of tails: dark matter interpretations of the fermi GeV excess in light of background model systematics. Phys. Rev. D **91**(6), 063003 (2015). arXiv:1411.4647 [hep-ph]
6. Ibe, M., Murayama, H., Yanagida, T.T.: Breit-Wigner enhancement of dark matter annihilation. Phys. Rev. D **79**, 095009 (2009). arXiv:0812.0072 [hep-ph]
7. Butter, A., Plehn, T., Rauch, M., Zerwas, D., Henrot-Versillé, S., Lafaye, R.: Invisible Higgs decays to Hooperons in the NMSSM. Phys. Rev. D **93**, 015011 (2016). arXiv:1507.02288 [hep-ph]
8. Bauer, M., Butter, A., Desai, N., Gonzalez-Fraile, J., Plehn, T.: Validity of dark matter effective theory. Phys. Rev. D **95**(7), 075036 (2017). arXiv:1611.09908 [hep-ph]

Chapter 6
Direct Searches

The experimental strategy for direct dark matter detection is based on measuring a recoil of a nucleus after scattering with WIMP dark matter. For this process we can choose the optimal nuclear target based on the largest possible recoil energy. We start with the non-relativistic relation between the momenta in relative coordinates between the nucleus and the WIMP, assuming a nucleus composed of A nucleons and with charge Z. The relative WIMP velocity $v_0/2$ is defined in Eq. (3.20), so in terms of the reduced mass $m_A m_\chi/(m_A + m_\chi)$ we find

$$2m_A E_A = |\vec{p}_A|^2 \approx \left(\frac{m_A m_\chi}{m_A + m_\chi}\right)^2 \frac{v_0^2}{4} \Leftrightarrow E_A = \frac{m_A}{(m_A + m_\chi)^2} m_\chi^2 \frac{v_0^2}{8}$$

$$\Rightarrow \frac{dE_A}{dm_A} = \left[\frac{1}{(m_A + m_\chi)^2} + \frac{(-2)m_A}{(m_A + m_\chi)^3}\right] m_\chi^2 \frac{v_0^2}{8} \overset{!}{=} 0 \Leftrightarrow \boxed{m_A = m_\chi}$$

$$\Rightarrow E_A = \frac{m_\chi^2}{4} \frac{1}{2m_\chi} \frac{v_0^2}{4} = \frac{m_\chi}{32} v_0^2 \approx 10^4 \text{ eV}, \tag{6.1}$$

with $v_0 \approx 1/1300$ and for a dark matter around 1 TeV. Because of the above relation, an experimental threshold from the lowest observable recoil can be directly translated into a lower limit on dark matter masses we can probe with such experiments. This also tells us that for direct detection all momentum transfers are very small compared to the electroweak or WIMP mass scale. Similar masses of WIMP and nuclear targets produce the largest recoil in the 10 keV range. Remembering that the Higgs mass in the Standard Model is roughly the same as the mass of the gold atom we know that it should be possible to find appropriate nuclei, for example Xenon with a nucleus including $A = 131$ nucleons, of which $Z = 54$ are protons.

© Springer Nature Switzerland AG 2019
M. Bauer, T. Plehn, *Yet Another Introduction to Dark Matter*,
Lecture Notes in Physics 959, https://doi.org/10.1007/978-3-030-16234-4_6

Strictly speaking, the dark matter velocity relevant for direct detection is a combination of the thermal, un-directional velocity $v_0 \approx 1/1300$ and the earth's movement around the sun,

$$v_{\text{earth-sun}}\, c = 15000 \cos\left(2\pi\, \frac{t - 152.5\,\text{d}}{365.25\,\text{d}}\right) \frac{\text{m}}{\text{s}}$$

$$\Leftrightarrow v_{\text{earth-sun}} = 5 \cdot 10^{-5} \cos\left(2\pi\, \frac{t - 152.5\,\text{d}}{365.25\,\text{d}}\right) \approx \frac{v_0}{15} \cos\left(2\pi\, \frac{t - 152.5\,\text{d}}{365.25\,\text{d}}\right).$$

$$(6.2)$$

If we had full control over all annual modulations in a direct detection scattering experiment we could use this modulation to confirm that events are indeed due to dark matter scattering.

Given that a dark matter particle will (typically) not be charged under $SU(3)_c$, the interaction of the WIMP with the partons inside the nucleons bound in the nucleus will have to be mediated by electroweak bosons or the Higgs. We expect a WIMP charged under $SU(2)_L$ to couple to a nucleus by directly coupling to the partons in the nucleons through Z-exchange. This means with increased resolution we have to compute the scattering processes for the nucleus, the nucleons, and the partons:

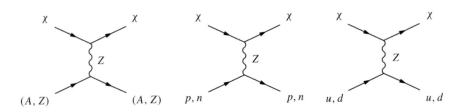

This gauge boson exchange will be dominated by the valence quarks in the combinations $p \approx (uud)$ and $n \approx (udd)$. Based on the interaction of individual nucleons, which we will calculate below, we can express the dark matter interaction with a heavy nucleus as

$$\sigma^{\text{SI}}(\chi A \to \chi A) = \frac{1}{16\pi s} \sum_N \overline{|Z\mathcal{M}_p + (A - Z)\mathcal{M}_n|^2}$$

$$= \begin{cases} \dfrac{A^2}{64\pi s} \overline{\sum_N |\mathcal{M}_p + \mathcal{M}_n|^2} & \text{for } Z = A/2 \\[3ex] \dfrac{A^2}{16\pi s} \overline{\sum_N |\mathcal{M}_n|^2} & \text{for } \mathcal{M}_p = \mathcal{M}_n. \end{cases}$$

$$(6.3)$$

We refer to this coherent interaction as spin-independent scattering with the cross section σ^{SI}. The scaling with A^2 appears as long as the exchange particle probes all nuclei in the heavy nucleon coherently, which means we have to square the sum of the individual matrix elements. The condition for coherent scattering can be formulated in terms of the size of the nucleus and the wavelength of the momentum transfer $\sqrt{2m_A E_A}$,

$$\sqrt{2m_A E_A} \approx \sqrt{2Am_p E_A} < \frac{1}{A^{1/3} r_p} \approx \frac{m_p}{A^{1/3}} \Leftrightarrow E_A < \frac{10^9 \text{ eV}}{2A^{5/3}} \cdot \quad (6.4)$$

for $m_p \approx 1$ GeV. This is clearly true for the typical recoils given in Eq. (6.1). In the next step, we need to compute the interaction to the individual A nucleons in terms of their partons. Because there are very different types of partons, valence quarks, sea quarks, and gluons, with different quantum numbers, this calculation is best described in a specific model.

One of the most interesting theoretical questions in direct detection is how different dark matter candidates couple to the non-relativistic nuclei. The general trick is to link the nucleon mass (operator) to the nucleon-WIMP interaction (operator). We know that three quarks can form a color singlet state; in addition, there will be a gluon and a sea quark content in the nucleons, but in a first attempt we assume that those will play a sub-leading role for the nucleon mass or its interaction to dark matter, as long as the mediator couples to the leading valence quarks. We start with the nucleon mass operator evaluated between two nucleon states and write it in terms of the partonic quark constituents,

$$\langle N | m_N \mathbb{1} | N \rangle = m_N \langle N | N \rangle = \sum_q \langle N | m_q \bar{q} q | N \rangle = \sum_q m_q \langle N | \bar{q} q | N \rangle , \quad (6.5)$$

assuming an appropriate definition of the constituent masses. Based on the same formalism we can write the nucleon–WIMP interaction operator in terms of the quark parton content,

$$\langle N | \sum_q \chi \chi \bar{q} q | N \rangle = \sum_q \chi \chi \langle N | \bar{q} q | N \rangle . \quad (6.6)$$

These two estimates suggest that we can link the nucleon interaction operator to the nucleon mass operator in the naive quark parton model. Based on the nucleon mass we define a non-relativistic quark density inside the nucleon as

$$f_N := \langle N | N \rangle \overset{\text{Eq. (6.5)}}{=} \sum_q \frac{m_q}{m_N} \langle N | \bar{q} q | N \rangle = \sum_q f_q \quad \Leftrightarrow \quad f_q := \frac{m_q}{m_N} \langle N | \bar{q} q | N \rangle$$

$$\Rightarrow \langle N | \sum_q \chi \chi \bar{q} q | N \rangle = \sum_q \chi \chi \frac{m_N}{m_q} f_q . \tag{6.7}$$

The form factors f_q describe the probability of finding a (valence) quark inside the proton or neutron at a momentum transfer well below the nucleon mass. They can for example be computed using lattice gauge theory.

6.1 Higgs Portal

The issue with Eq. (6.7) is that it neither includes gluons nor any quantum effects. Things become more interesting with a Higgs-mediated WIMP–nucleon interaction, as we encounter it in our Higgs portal models. To cover this case we need to compute both, the nucleon mass and the WIMP–nucleon interaction operators beyond the quark parton level. From LHC we know that at least for relativistic protons the dominant Higgs coupling is through the gluon content. In the Standard Model the Higgs coupling to gluons is mediated by a top loop, which does not decouple for large top masses. The fact that, in contrast, the top quark does decouple from the nucleon mass will give us a non-trivial form factor for gluons.

Defining our quantum field theory framework, in proper QCD two terms contribute to the nucleon mass: the valence quark masses accounted for in Eq. (6.5) and the strong interaction, or gluons, leading to a binding energy. This view is supported by the fact that pions, consisting of two quarks, are almost an order of magnitude lighter than protons and neutrons, with three quarks. We can describe both sources of the nucleon mass using the energy–momentum tensor $T^{\mu\nu}$ as it appears for example in the Einstein–Hilbert action in Eq. (1.15),

$$\boxed{m_N \langle N|N \rangle = \langle N|T^\mu_\mu|N \rangle} \, . \tag{6.8}$$

Scale invariance, or the lack of fundamental mass scales in our theory implies that the energy–momentum tensor is traceless. A non-zero trace of the energy–momentum tensor indicates a change in the Lagrangian with respect to a scale variation, where in our units a variation of the length scale and a variation of the energy scale are equivalent. Lagrangians which are symmetric under such a scale variation cannot include explicit mass terms, because those correspond to a fixed energy scale.

In addition to the quark masses, for the general form of the nucleon mass given in Eq. (6.8) we need to consider contributions from the running strong coupling to the trace of the energy–momentum tensor. At one-loop order the running of α_s with the underlying energy scale p^2 is given by

$$\alpha_s(p^2) = \frac{1}{b_0 \log \dfrac{p^2}{\Lambda_{\text{QCD}}^2}} \quad \text{with} \quad b_0 = -\frac{1}{4\pi}\left(\frac{2n_q}{3} - \frac{11}{3} N_c\right), \tag{6.9}$$

and an appropriate reference value of $\Lambda_{\text{QCD}} \approx 200\,\text{MeV}$. Mathematically, such a reference mass scale has to appear in any problem which involves a logarithmic running, i.e. which would otherwise force us to take the logarithm of a dimensionful scale p^2. Physically, this scale is defined by the point at which the strong coupling explodes and we need to switch degrees of freedom. That occurs at positive energy scales as long as $b_0 > 0$, or as long as the gluons dominate the running of α_s. Because the running of the strong coupling turns the dimensionless parameter α_s into the dimensionful parameter Λ_{QCD}, this mechanism is called dimensional transmutation.

The contribution of the running strong coupling to the nucleon mass is given through the kinetic gluon term in the Lagrangian, combined with the momentum variation of the strong coupling. Altogether we find

$$
\begin{aligned}
m_N \langle N|N \rangle &= \sum_q m_q \langle N|\bar{q}q|N \rangle + \frac{2}{\alpha_s} \frac{d\alpha_s}{d\log p^2} \langle N|G_{\mu\nu}^a G^{a\,\mu\nu}|N \rangle \\
&= \sum_q m_q \langle N|\bar{q}q|N \rangle - \frac{\alpha_s b_0}{2} \langle N|G_{\mu\nu}^a G^{a\,\mu\nu}|N \rangle \\
&= \sum_q m_q \langle N|\bar{q}q|N \rangle + \frac{\alpha_s}{8\pi} \left(\frac{2n_q}{3} - \frac{11}{3} N_c \right) \langle N|G_{\mu\nu}^a G^{a\,\mu\nu}|N \rangle ,
\end{aligned}
$$
(6.10)

again written at one loop and neglecting the anomalous dimension of the quark fields. One complication in this formula is the appearance of all six quark fields in the sum, suggesting that all quarks contribute to the nucleon mass. While this is true for the up and down valence masses, and possibly for the strange mass, the three heavier quarks hardly appear in the nucleon. Instead, they contribute to the nucleon mass through gluon splitting or self energy diagrams in the gluon propagator. We can compute this contribution in terms of a heavy quark effective theory, giving us the leading contribution per heavy quark

$$
\langle N|\bar{q}q|N \rangle \bigg|_{c,b,t} = -\frac{\alpha_s}{12\pi m_q} \langle N|G_{\mu\nu}^a G^{a\,\mu\nu}|N \rangle + \mathcal{O}\left(\frac{1}{m_q^3} \right) .
$$
(6.11)

We can insert this result in the above expression and find the complete expression for the nucleon mass operator

$$
\begin{aligned}
m_N \langle N|N \rangle &= \sum_{u,d,s} m_q \langle N|\bar{q}q|N \rangle - \sum_{c,b,t} \frac{\alpha_s}{12\pi} \langle N|G_{\mu\nu}^a G^{a\,\mu\nu}|N \rangle \\
&\quad + \frac{\alpha_s}{8\pi} \left(\frac{2 \times 6}{3} - \frac{11}{3} N_c \right) \langle N|G_{\mu\nu}^a G^{a\,\mu\nu}|N \rangle \\
&= \sum_{u,d,s} m_q \langle N|\bar{q}q|N \rangle + \frac{\alpha_s}{8\pi} \left(\frac{2 \times 3}{3} - \frac{11}{3} N_c \right) \langle N|G_{\mu\nu}^a G^{a\,\mu\nu}|N \rangle .
\end{aligned}
$$
(6.12)

Starting from the full beta function of the strong coupling this result implies that we only need to consider the running due to the three light-flavor quarks and the gluon itself for the nucleon mass prediction,

$$m_N \langle N|N \rangle = \sum_{u,d,s} m_q \langle N|\bar{q}q|N \rangle - \frac{\alpha_s b_0^{(u,d,s)}}{2} \langle N|G_{\mu\nu}^a G^{a\,\mu\nu}|N \rangle$$

$$\text{with} \quad b_0^{(u,d,s)} = -\frac{1}{4\pi} \left(\frac{2n_{\text{light}}}{3} - \frac{11}{3}N_c \right) . \qquad (6.13)$$

This reflects a full decoupling of the heavy quarks in their contribution to the nucleon mass. From the derivation it is clear that the same structure appears for any number of light quarks defining our theory.

Exactly in the same way we now describe the WIMP–nucleon interaction in terms of six quark flavors. The light quarks, including the strange quark, form the actual quark content of the nucleon. Virtual heavy quarks occur through gluon splitting at the one-loop level. In addition to the small Yukawa couplings of the light quarks we know from LHC physics that we can translate the Higgs-top interaction into an effective Higgs–gluon interaction. In the limit of large quark masses the loop-induced coupling defined by the Feynman diagram

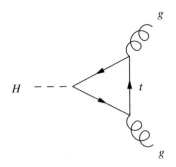

is given by

$$\mathscr{L}_{ggH} \supset \frac{1}{v_H} g_{ggH} \, H \, G^{\mu\nu} G_{\mu\nu} \qquad \text{with} \quad g_{ggH} = \frac{\alpha_s}{12\pi} . \qquad (6.14)$$

In terms of an effective field theory the dimension-5 operator scales like $1/v$ and not $1/m_t$. The reason is that the dependence on the top mass in the loop and on the Yukawa coupling in the numerator cancel exactly in the limit of small momentum transfer through the Higgs propagator. Unlike for the nucleon mass operators this means that in the Higgs interaction the Yukawa coupling induces a non-decoupling feature in our theory. Using this effective field theory level we can successively compute the ggH^{n+1} coupling from the ggH^n coupling via

$$g_{ggH^{n+1}} = m_q^{n+1} \frac{\partial}{\partial m_q} \left(\frac{1}{m_q^n} g_{ggH^n} \right).$$

(6.15)

This relation also holds for $n = 0$, which means it formally links the Higgs–nucleon coupling operator to the nucleon mass operator in Eq. (6.13). The only difference between the effective Higgs-gluon interaction at LHC energies and at direct detection energies is that in direct detection all three quarks c, b, t contribute to the effective interaction defined in Eq. (6.14).

Keeping this link in mind we see that the Higgs-mediated WIMP interaction operator again consists of two terms

$$\langle N | \sum_{u,d,s} m_q H \bar{q} q | N \rangle - \langle N | \sum_{c,b,t} \frac{\alpha_s}{12\pi} H G^a_{\mu\nu} G^{a\,\mu\nu} | N \rangle.$$

(6.16)

The corresponding Feynman diagrams at the parton level are:

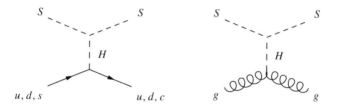

The Yukawa interaction, described by the first terms in Eq. (6.16) has a form similar to the nucleon mass in Eq. (6.13). Comparing the two formulas for light quarks only we indeed find

$$\langle N | \sum_{q} m_q H \bar{q} q | N \rangle \Big|_{u,d,s} = H \sum_{u,d,s} m_q \langle N | \bar{q} q | N \rangle \stackrel{\text{Eq. (6.13)}}{=} H m_N \langle N | N \rangle \Big|_{u,d,s}.$$

(6.17)

This reproduces the simple recipe for computing the light-quark-induced WIMP–nucleon interaction as proportional to the nucleon mass. The remaining, numerically dominant gluonic terms is defined in the so-called chiral limit $m_{u,d,s} = 0$. Because of the non-decoupling behavior this contribution is independent of the heavy quark mass, so we find for n_{heavy} heavy quarks

$$\langle N|\sum_q m_q H\bar{q}q|N\rangle\bigg|_{c,b,t} \overset{\text{Eq.\,(6.16)}}{=} -\frac{2n_{\text{heavy}}}{3}\frac{\alpha_s}{8\pi}H\,\langle N|G_{\mu\nu}^a G^{a\,\mu\nu}|N\rangle$$

$$\neq -\frac{\alpha_s}{8\pi}\left(\frac{11}{3}N_c - \frac{2n_{\text{light}}}{3}\right)H\,\langle N|G_{\mu\nu}^a G^{a\,\mu\nu}|N\rangle \overset{\text{Eq.\,(6.13)}}{=} Hm_N\langle N|N\rangle\bigg|_{c,b,t,g}.$$

$$(6.18)$$

The contribution to the nucleon mass comes from the gluon and n_{light} light quark loops, while the gluonic contribution to the nucleon–Higgs coupling is driven by the n_{heavy} heavy quark loops. The boundary condition is $n_{\text{light}} + n_{\text{heavy}} = 6$. At the energy scale of direct detection we can compensate for this mismatch in the Higgs–nucleon coupling of the naive scaling between the nucleon mass and nucleon Yukawa interaction shown in Eq. (6.7). We simply include an additional factor

$$\sum_q \frac{m_q}{m_N}\langle N|H\bar{q}q|N\rangle\bigg|_{c,b,t} = \frac{\dfrac{2n_{\text{heavy}}}{3}}{\dfrac{11}{3}N_c - \dfrac{2n_{\text{light}}}{3}}H\,\langle N|N\rangle\bigg|_{c,b,t,g}, \qquad (6.19)$$

which we can estimate at leading order and at energy scales relevant for direct dark matter searches to be

$$\frac{\dfrac{2n_{\text{heavy}}}{3}}{\dfrac{11}{3}N_c - \dfrac{2n_{\text{light}}}{3}} \overset{n_{\text{light}}=3}{=} \frac{3\times\dfrac{2}{3}}{11 - \dfrac{2\times 3}{3}} = \frac{2}{9}. \qquad (6.20)$$

This effect leads to a suppression of the already small Higgs–nucleon interaction at low momentum transfer. The exact size of the suppression depends on the number of active light quarks in our effective theory, which in turn depends on the momentum transfer.

At the parton level, the weakly interacting part of the calculation of the nucleon–WIMP scattering rate closely follows the calculation of WIMP annihilation in Eq. (4.7). In the case of direct detection the valence quarks in the nucleons couple through a t-channel Higgs to the dark matter scalar S. We account for the parton nature of the three relevant heavy quarks by writing the nucleon Yukawa coupling as $f_N m_N \times 2/9$,

$$\mathcal{M} = \bar{u}(k_2)\frac{-2if_N m_N}{9v_H}u(k_1)\frac{-i}{(k_1-k_2)^2 - m_H^2}(-2i\lambda_3 v_H). \qquad (6.21)$$

For an incoming and outgoing fermion the two spinors are $\bar{u}$ and u. As long as the Yukawa coupling is dominated by the heavy quarks, it will be the same for neutrons and protons, i.e. $\mathcal{M}_p = \mathcal{M}_n$. We have to square this matrix element, paying attention to the spinors v and u, and then sum over the spins of the external fermions. In this case we already know that we are only interested in scattering in the low-energy limit, i.e. $|(k_1 - k_2)^2| \ll m_N^2 \ll m_H^2$,

$$\sum_{\text{spin}} |\mathcal{M}|^2$$

$$= \frac{16}{81} \lambda_3^2 f_N^2 m_N^2 \left(\sum_{\text{spin}} u(k_2) \bar{u}(k_2) \right) \left(\sum_{\text{spin}} u(k_1) \bar{u}(k_1) \right) \frac{1}{\left[(k_1 - k_2)^2 - m_H^2 \right]^2}$$

$$= \frac{16}{81} \lambda_3^2 f_N^2 m_N^2 \ \text{Tr} \left[(\not{k}_2 + m_N \mathbb{1}) (\not{k}_1 + m_N \mathbb{1}) \right] \frac{1}{\left[(k_1 - k_2)^2 - m_H^2 \right]^2}$$

$$= \frac{32}{81} \lambda_3^2 f_N^2 m_N^2 \left[2 k_1 \cdot k_2 + 2 m_N^2 \right] \frac{1}{\left[(k_1 - k_2)^2 - 2 m_H^2 \right]^2}$$

$$= \frac{32}{81} \lambda_3^2 f_N^2 m_N^2 \left[-(k_1 - k_2)^2 + 4 m_N^2 \right] \frac{1}{\left[(k_1 - k_2)^2 - m_H^2 \right]^2}$$

$$\approx \frac{128}{81} \lambda_3^2 f_N^2 \frac{m_N^4}{m_H^4} \Rightarrow \overline{\sum_{\text{spin,color}} |\mathcal{M}|^2} = \frac{64}{81} \lambda_3^2 f_N^2 \frac{m_N^4}{m_H^4} \qquad (6.22)$$

The cross section in the low-energy limit is by definition spin-independent and becomes

$$\sigma^{\text{SI}}(SN \to SN) = \frac{1}{16 \pi s} \overline{\sum} |\mathcal{M}|^2$$

$$= \frac{1}{16 \pi (m_S + m_N)^2} \frac{64}{81} \lambda_3^2 f_N^2 \frac{m_N^4}{m_H^4} \approx \frac{4 \lambda_3^2 f_N^2}{81 \pi} \frac{m_N^4}{m_H^4} \frac{1}{m_S^2},$$

$$\qquad (6.23)$$

where in the last step we assume $m_S \gg m_N$. For WIMP–Xenon scattering this gives us

$$\boxed{ \sigma^{\text{SI}}(SA \to SA) = \frac{4 \lambda_3^2 f_N^2 A^2}{81 \pi} \frac{m_N^4}{m_H^4} \frac{1}{m_S^2} = 6 \cdot 10^{-7} \frac{\lambda_3^2}{m_S^2} . } \qquad (6.24)$$

The two key ingredients to this expression can be easily understood: the suppression $1/m_H^4$ appears after we effectively integrate out the Higgs in the t-channel, and the high power of m_N^4 occurs because in the low-energy limit the Higgs coupling to fermions involve a chirality flip and hence one power of m_N for each coupling. The angle-independent matrix element in the low-energy limit can easily be translated into a spectrum of the scattering angle, which will then give us the recoil spectrum, if desired. We limit ourselves to the total rate, assuming that the appropriate WIMP mass range ensures that the total cross section gets converted into measurable recoil. This approach reflects the fact that we consider the kinematics of scattering processes and hence the existence of phase space a topic for experimental lectures.

Next, we can ask which range of Higgs portal parameters with the correct relic density, as shown in Fig. 4.1, is accessible to direct detection experiments. According to Eq. (6.24) the corresponding cross section first becomes small when $\lambda_3 \ll 1$, which means $m_S \lesssim m_H/2$ with the possibility of explaining the Fermi galactic center excess. Second, the direct detection cross section is suppressed for heavy dark matter and leads to a scaling $\lambda_3 \propto m_S$.

From Eq. (4.21) we know that a constant annihilation rate leading to the correct relic density also corresponds to $\lambda_3 \propto m_S$. However, while the direct detection rate features an additional suppression through the nucleon mass m_N, the annihilation rate benefits from several subleading annihilation channels, like for example the annihilation to two gauge bosons or two top quarks. This suggests that for large m_S the two lines of constant cross sections in the λ_3-m_S plane run almost in parallel, with a slightly smaller slope for the annihilation rate. This is exactly what we observe in Fig. 4.1, leaving heavy Higgs portal dark matter with $m_S \gtrsim 300\,\mathrm{GeV}$ a viable model for all observations related to cold dark matter. This minimal dark matter mass constraint rapidly increases with new direct detection experiments coming online. On the other hand, from our discussion of the threshold behavior in Sect. 3.5 is should be clear that we can effectively switch off all direct detection constraints by making the scalar Higgs mediator a pseudo-scalar.

Finally, we can modify our model and the quantitative link between the relic density and direct detection, as illustrated in Fig. 4.1. The typical renormalizable Higgs portal includes a scalar dark matter candidate. However, if we are willing to include higher-dimensional terms in the Lagrangian we can combine the Higgs portal with fermionic and vector dark matter. This is interesting in view of the velocity dependence discussed in Sect. 3.4. The annihilation of dark matter fermions is velocity-suppressed at threshold, so larger dark matter couplings predict the observed relic density. Because direct detection is not sensitive to the annihilation threshold, it will be able to rule out even the mass peak region for fermionic dark matter (Fig. 6.1).

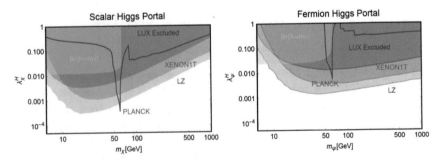

Fig. 6.1 Relic density (labelled PLANCK) vs direct dark matter detection constraints. The dark matter agent is switched from a real scalar (left) to a fermion (right). Figure from Ref. [1]

6.2 Supersymmetric Neutralinos

In supersymmetry with its fermionic dark matter candidate, nucleon–neutralino scattering is described by four-fermion operators, just like in Fermi's theory. The reason is that all intermediate particles coupling the two neutralinos to two quarks are far below their mass shell. Accounting for the mass dimension through a scalar mediator mass scale $\Lambda \approx m_{h^0}$, the matrix element reads

$$\mathcal{M} = \frac{g_{NN\tilde{\chi}_1^0\tilde{\chi}_1^0}}{\Lambda^2} \, \bar{v}_{\tilde{\chi}_1^0} v_{\tilde{\chi}_1^0} \, \bar{u}_N u_N$$

$$\sum_{\text{spins}} |\mathcal{M}|^2 = \frac{g_{NN\tilde{\chi}_1^0\tilde{\chi}_1^0}^2}{\Lambda^4} \, \text{Tr}\left[(\not{p}_2 - m_{\tilde{\chi}_1^0}\mathbb{1})\,(\not{p}_1 - m_{\tilde{\chi}_1^0}\mathbb{1}) \right]$$

$$\times \text{Tr}\left[(\not{k}_2 + m_N\mathbb{1})\,(\not{k}_1 + m_N\mathbb{1}) \right]$$

$$\approx 64 g_{NN\tilde{\chi}_1^0\tilde{\chi}_1^0}^2 \frac{m_{\tilde{\chi}_1^0}^2 m_N^2}{\Lambda^4} \Rightarrow \sum_{\text{spin,color}} |\mathcal{M}|^2 = 16 \, g_{NN\tilde{\chi}_1^0\tilde{\chi}_1^0}^2 \frac{m_{\tilde{\chi}_1^0}^2 m_N^2}{\Lambda^4}.$$

$$(6.25)$$

The corresponding spin-independent cross section mediated by the Standard Model Higgs in the low-energy limit is then

$$\boxed{\sigma^{\text{SI}}(\tilde{\chi}_1^0 N \to \tilde{\chi}_1^0 N) \approx \frac{g_{NN\tilde{\chi}_1^0\tilde{\chi}_1^0}^2}{\pi} \frac{m_N^2}{m_{h^0}^4}.} \qquad (6.26)$$

As for the Higgs portal case in Eq. (6.24) the rate is suppressed by the mediator mass to the fourth power. The lower power of m_N^2 appears only because we absorb the Yukawa coupling in $g_{NN\tilde{\chi}_1^0\tilde{\chi}_1^0} = 2m_N f_N/9$,

$$g_{NN\tilde{\chi}_1^0\tilde{\chi}_1^0} = \frac{g_{h^0\tilde{\chi}_1^0\tilde{\chi}_1^0} \, 2f_N m_N}{9}$$

$$\propto \frac{2f_N m_N}{9} \, (g' N_{11} - g N_{12}) \, (\sin\alpha \, N_{13} + \cos\alpha \, N_{14}) \,, \qquad (6.27)$$

following Eq. (4.43). We see that this scaling is identical to the Higgs portal case in Eq. (6.24), but with an additional suppression through the difference in mixing angles in the neutralino and Higgs sectors.

However, in supersymmetric models the dark matter mediator will often be the Z-boson, because the interaction $g_{NN\tilde{\chi}_1^0\tilde{\chi}_1^0}$ is not suppressed by a factor of the kind m_N/v. In this case we need to describe a (transverse) vector coupling between the WIMP and the nucleon in our four-fermion interaction. Following exactly the argument as for the scalar exchange we can look at a Z-mediated interaction between a (Dirac) fermion χ and the nucleons,

$$\mathcal{M} = \frac{g_{NN\chi\chi}}{\Lambda^2} \, \bar{v}_\chi \gamma_\mu v_\chi \, \bar{u}_N \gamma^\mu u_N$$

$$\sum_{\text{spins}} |\mathcal{M}|^2 = \frac{g_{NN\chi\chi}^2}{\Lambda^4} \, \text{Tr}\left[(\not{p}_2 - m_\chi)\gamma_\mu (\not{p}_1 - m_\chi)\gamma_\nu \right]$$

$$\times \text{Tr}\left[(\not{k}_2 + m_N)\gamma^\mu \, (\not{k}_1 + m_N)\gamma^\nu \right]$$

$$\approx \frac{g_{NN\chi\chi}^2}{\Lambda^4} \, (8m_\chi^2)(8m_N^2)$$

$$\approx 64 g_{NN\chi\chi}^2 \, \frac{m_\chi^2 m_N^2}{\Lambda^4} \qquad \Rightarrow \qquad \boxed{\sigma^{\text{SI}}(\chi N \to \chi N) \approx \frac{4 g_{NN\chi\chi}^2}{\pi} \, \frac{m_N^2}{\Lambda^4}}.$$

$$(6.28)$$

The spin-independent cross section mediated by a gauge boson is typically several orders of magnitude larger than the cross section mediated by Higgs exchange. This means that models with dark matter fermions coupling to the Z-boson will be in conflict with direct detection constraints. For the entire relic neutralino surface with pure and mixed states the spin-independent cross sections are shown in Fig. 6.2. The corresponding current and future exclusion limits are indicated in Fig. 6.3. The so-called neutrino floor, which can be reached within the next decade, is the parameter region where the expected neutrino background will make standard direct detection searches more challenging.

The problem with this result in Eq. (6.28) is that it does not hold for Majorana fermions, like the neutralino $\tilde{\chi}_1^0$. From the discussion in Sect. 3.5 we know that a vector mediator cannot couple Majorana fermions to the nucleus, so we are left with the corresponding axial vector exchange,

$$\mathcal{M} = \frac{g_{NN\tilde{\chi}_1^0\tilde{\chi}_1^0}}{\Lambda^2} \, \bar{v}_{\tilde{\chi}_1^0}\gamma_\mu\gamma_5 v_{\tilde{\chi}_1^0} \, \bar{u}_N \gamma^\mu \gamma_5 u_N; \, ,$$

$$\sum_{\text{spins}} |\mathcal{M}|^2 = \frac{g_{NN\tilde{\chi}_1^0\tilde{\chi}_1^0}^2}{\Lambda^4} \, \text{Tr}\left[(\not{p}_2 - m_{\tilde{\chi}_1^0}) \, \gamma_5\gamma_\mu \, (\not{p}_1 - m_{\tilde{\chi}_1^0}) \, \gamma_5\gamma_\nu \right]$$

$$\text{Tr}\left[(\not{k}_2 + m_N) \, \gamma_5\gamma^\mu \, (\not{k}_1 + m_N) \, \gamma_5\gamma^\nu \right]. \qquad (6.29)$$

For axial vector couplings the current is defined by $\gamma_\mu\gamma_5$. This means it depends on the chirality or the helicity of the fermions. The spin operator is defined in terms of the Dirac matrices as $\vec{s} = \gamma_5\gamma^0\vec{\gamma}$. This indicates that the axial vector coupling

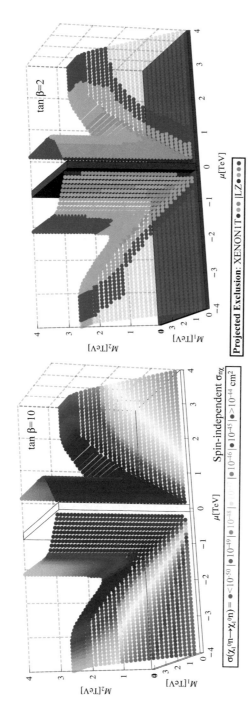

Fig. 6.2 Left: spin-independent nucleon-scattering cross-section for relic neutralinos. Right: relic neutralino exclusions from XENON100 and LUX and prospects from XENON1T and LZ. The boxed out area denotes the LEP exclusion. Figure from Ref. [2]

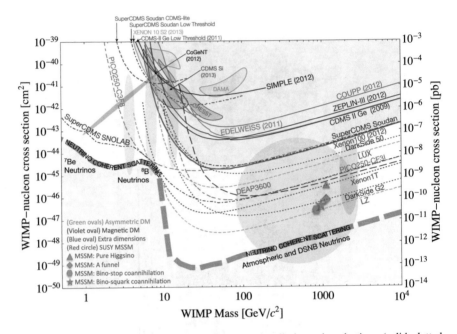

Fig. 6.3 Spin-independent WIMP–nucleon cross section limits and projections (solid, dotted, dashed curves) and hints for WIMP signals (shaded contours) and projections (dot and dot-dashed curves) for direct detection experiments. The yellow region indicates dangerous backgrounds from solar, atmospheric, and diffuse supernovae neutrinos. Figure from Ref. [3]

actually is a coupling to the spin of the nucleon. This is why the result is called a spin-dependent cross section, which for each nucleon reads

$$\sum_{spins} |\mathcal{M}|^2 \approx 16 \times 4g_{NN\tilde{\chi}_1^0\tilde{\chi}_1^0}^2 \frac{m_{\tilde{\chi}_1^0}^2 m_N^2}{\Lambda^4}$$

$$\Rightarrow \quad \boxed{\sigma^{SD}(\tilde{\chi}_1^0 N \to \tilde{\chi}_1^0 N) \approx \frac{4g_{NN\tilde{\chi}_1^0\tilde{\chi}_1^0}^2}{\pi} \frac{m_N^2}{\Lambda^4}} . \quad (6.30)$$

Again, we can read off Eq. (4.43) that for the light quarks $q = u, d, s$ the effective coupling should have the form

$$g_{NN\tilde{\chi}_1^0\tilde{\chi}_1^0} = g_{Z\tilde{\chi}_1^0\tilde{\chi}_1^0} g_{Zqq} \propto g^2 \left(|N_{13}|^2 - |N_{14}|^2 \right) . \quad (6.31)$$

The main difference between the spin-independent and spin-dependent scattering is that for the coupling to the nucleon spin we cannot assume that the couplings to all nucleons inside a nucleus add coherently. Instead, we need to link the spin representations of the nucleus to spin representations of each nucleon. Instead of finding a coherent enhancement with Z^2 or A^2 the result scales like A, weighted by Clebsch-Gordan coefficients which appear from reducing out the combination of the spin-1/2 nucleons. Nevertheless, direct detection strongly constrains the higgsino content of the relic neutralino, with the exception of a pure higgsino, where the two terms in $g_{Z\tilde{\chi}_1^0\tilde{\chi}_1^0}$ cancel each other.

References

1. Arcadi, G., Dutra, M., Ghosh, P., Lindner, M., Mambrini, Y., Pierre, M., Profumo, S., Queiroz, F.S.: The waning of the WIMP? A review of models, searches, and constraints. Eur. Phys. J. C **78**(3), 203 (2018). arXiv:1703.07364 [hep-ph]
2. Bramante, J., Desai, N., Fox, P., Martin, A., Ostdiek, B., Plehn, T.: Towards the final word on neutralino dark matter. Phys. Rev. D **93**(6), 063525 (2016). arXiv:1510.03460 [hep-ph]
3. Feng, J.L., et al.: Planning the Future of U.S. Particle Physics (Snowmass 2013): Chapter 4: Cosmic Frontier. arXiv:1401.6085 [hep-ex]

Chapter 7
Collider Searches

Collider searches for dark matter rely on two properties of the dark matter particle: first, the new particles have to couple to the Standard Model. This can be either a direct coupling for example to the colliding leptons and quarks, or an indirect coupling through an mediator. Second, we need to measure traces of particles which interact with the detectors as weakly as for example neutrinos do. And unlike dedicated neutrino detectors their collider counter parts do not include hundreds of cubic meters of interaction material. Under those boundary conditions collider searches for dark matter particles will benefit from several advantages:

1. we know the kinematic configuration of the dark matter production process. This is linked to the fact that most collider detectors are so-called multi-purpose detectors which can measure a great number of observables;
2. the large number of collisions (parametrized by the luminosity $\mathscr{L}$) can give us a large number of dark matter particles to analyze. This allows us to for example measure kinematic distributions which reflect the properties of the dark matter particle;
3. all background processes and all systematic uncertainties can be studied, understood, and simulated in detail. Once an observation of a dark matter particle passes all conditions the collider experiments require for a discovery, we will know that we discovered such a new particle. Otherwise, if an anomaly turns out to not pass these conditions we have at least in my life time always been able to identify what the problem was.

One weakness we should always keep in mind is that a particle which does not decay while crossing the detector and which interacts weakly enough to not leave a trace does not have to be stable on cosmological time scales. To make this statement we need to measure enough properties of the dark matter particle to for example predict its relic density the way we discuss it in Chap. 3.

© Springer Nature Switzerland AG 2019
M. Bauer, T. Plehn, *Yet Another Introduction to Dark Matter*,
Lecture Notes in Physics 959, https://doi.org/10.1007/978-3-030-16234-4_7

7.1 Lepton Colliders

The key observable we can compute and analyze at colliders is the number of events expected for a certain production and decay process in a given time interval. The number of events is the product of the luminosity $\mathcal{L}$ measured for example in inverse femtobarns, the total production cross section measured in femtobarns, and the detection efficiency measured in per-cent,[1]

$$N_{\text{events}} = \sigma_{\text{tot}} \, \mathcal{L} \, \Pi_j \epsilon_j \, . \tag{7.1}$$

This way the event rate is split into a collider-specific number describing the initial state, a process-specific number describing the physical process, and a detector-specific efficiency for each final state particle. The efficiency includes for example phase-space dependent cuts defining the regions of sensitivity of a given experiment, as well as the so-called trigger requirements defining which events are saved and looked at. This structure holds for every collider.

When it comes to particles with electroweak interactions the most influential experiments were ALEPH, OPAL, DELPHI, and L3 at the Large Electron-Positron Collider (LEP) at CERN. It ran from 1989 until 2000, first with a e^+e^- energy right on the Z pole, and then with energies up to 209 GeV. Its life-time integrated luminosity is 1 fb^{-1}. The results form running on the Z pole are easily summarized: the $SU(2)_L$ gauge sector shows no hints for deviations from the Standard Model predictions. Most of these results are based on an analysis of the Breit–Wigner propagator of the Z boson which we introduce in Eq. (4.11),

$$\sigma(e^+e^- \to Z) \propto \frac{E_{e^+e^-}^2}{(E_{e^+e^-}^2 - m_Z^2)^2 + m_Z^2 \Gamma_Z^2} \, . \tag{7.2}$$

If we know what the energy of the incoming e^+e^- system is we can plot the cross section as a function of $E_{e^+e^-}$ and measure the Z mass and the Z width,

$$m_Z = (91.19 \pm 0.003)\,\text{GeV} \, , \qquad\qquad \Gamma_Z = (2.49 \pm 0.004)\,\text{GeV} \, . \tag{7.3}$$

From this Z mass measurement in relation to the W mass and the vacuum expectation value $v_H = 246$ GeV we can extract the top quark and Higgs masses, because these particles contribute to quantum corrections of the Z properties. The total Z width includes a partial width from the decay $Z \to \nu\bar{\nu}$, with a branching ratio around 20%. It comes from three generations of light neutrinos and is much larger than for example the 3.4% branching ratio of the decay $Z \to e^+e^-$. Under the assumption that only neutrinos contribute to the invisible Z decays we can translate the measurement of the partial width into a measurement of the number

---

[1]Cross sections and luminosities are two of the few observables which we do not measure in eV.

of light neutrinos, giving 2.98 ± 0.008. Alternatively, we can assume that there are three light neutrinos and use this measurement to constrain light dark matter with couplings to the Z that would lead to an on-shell decay, for example $Z \to \tilde{\chi}_1^0 \tilde{\chi}_1^0$ in our supersymmetric model. If a dark matter candidate relies on its electroweak couplings to annihilate to the observed relic density, this limit means that any WIMP has be heavier than

$$m_{\tilde{\chi}_1^0,s} > \frac{m_Z}{2} = 45\,\text{GeV} . \tag{7.4}$$

The results from the higher-energy runs are equally simple: there is no sign of new particles which could be singly or pair-produced in e^+e^- collisions. The Feynman diagram for the production of a pair of new particles, which could be dark matter particles, is

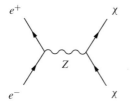

The experimental results mean that it is very hard to postulate new particles which couple to the Z boson or to the photon. The Feynman rules for the corresponding $f\bar{f}Z$ and $f\bar{f}\gamma$ couplings are

$$-i\gamma^\mu \left(\ell \mathbb{P}_L + r \mathbb{P}_R\right) \quad \text{with} \quad \ell = \frac{e}{s_w c_w} \left(T_3 - 2Qs_w^2\right) \quad r = \ell\Big|_{T_3=0} \quad (Zf\bar{f})$$

$$\ell = r = Qe \qquad\qquad (\gamma f\bar{f}) , \tag{7.5}$$

with the isospin quantum number $T_3 = \pm 1/2$ and $s_w^2 \approx 1/4$. Obviously, a pair of charged fermions will always be produced through an s-channel photon. If a particle has $SU(2)_L$ quantum numbers, the Z-coupling can be cancelled with the help of the electric charge, which leads to photon-induced pair production. Dark matter particles cannot be charged electrically, so for WIMPs there will exist a production process with a Z-boson in the s-channel. This result is important for co-annihilation in a more complex dark matter sector. For example in our supersymmetric model the charginos couple to photons, which means that they have to be heavier than

$$m_{\tilde{\chi}_1^\pm} > \frac{E_{e^+e^-}^{\max}}{2} = 104.5\,\text{GeV}; , \tag{7.6}$$

in order to escape LEP constraints. The problem of producing and detecting a pair of dark matter particles at any collider is that if we do not produce anything else those events with 'nothing visible happening' are hard to identify. Lepton colliders have one big advantage over hadron colliders, as we will see later: we know the kinematics of the initial state. This means that if, for example, we produce one invisibly decaying particle we can reconstruct its four-momentum from the initial state momenta and the final-state recoil momenta. We can then check whether for the majority of events the on-shell condition $p^2 = m^2$ with a certain mass is fulfilled. This is how OPAL managed to extract limits on Higgs production in the process $e^+e^- \rightarrow ZH$ without making any assumptions about the Higgs decay, notably including a decay to two invisible states. Unfortunately, because this analysis did not reach the observed Higgs mass of 126 GeV, it does not constrain our dark matter candidates in Higgs decays.

The pair production process

$$e^+e^- \rightarrow \gamma^* Z^* \rightarrow \chi\chi \tag{7.7}$$

is hard to extract experimentally, because we cannot distinguish it from an electron and positron just missing each other. The way out is to produce another particle in association with the dark matter particles, for example a photon with sufficiently large transverse momentum p_T

$$e^+e^- \rightarrow \tilde{\chi}_1^0 \tilde{\chi}_1^0 \gamma, \; SS\gamma \;, \tag{7.8}$$

Experimentally, this photon recoils against the two dark matter candidates, defining the signature as a photon plus missing momentum. A Feynman diagram for the production of a pair of dark matter particles and a photon through a Z-mediator is

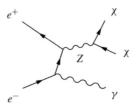

Because the photon can only be radiated off the incoming electrons, this process is often referred to as initial state radiation (ISR). Reconstructing the four-momentum of the photon allows us to also reconstruct the four-momentum of the pair of dark matter particles. The disadvantage is that a hard photon is only present in a small fraction of all e^+e^- collisions for example at LEP. This is one of the few instances where the luminosity or the size of the cross section makes a difference at LEP. Normally, the relatively clean e^+e^- environment allows us to build very efficient and very precise detectors, which altogether allows us to separate a signal from a usually small background cleanly. For example, the chargino mass limit in Eq. (7.6) applies to a wide set of new particles which decay into leptons and missing energy and is hard to avoid.

We should mention that for a long time people have discussed building another e^+e^- collider. Searching for new particles with electroweak interactions is one of the main motivations. Proposals range from a circular Higgs factory with limited energy due to energy loss in synchrotron radiation (FCC-ee/CERN or CEPC/China) to a linear collider with an energy up to 1 TeV (ILC/Japan), to a multi-TeV linear collider with a driving beam technology (CLIC/CERN).

7.2 Hadron Colliders and Mono-X

Historically, hadron colliders have had great success in discovering new, massive particles. This included UA1 and UA2 at SPS/CERN discovering the W and Z bosons, CDF and D0 at the Tevatron/Fermilab discovering the top quark, and most recently ATLAS and CMS at the LHC with their Higgs discovery. The simple reason is that protons are much heavier than electrons, which makes it easier to store large amounts of kinetic energy and release them in a collision. On the other hand, hadron collider physics is much harder than lepton collider physics, because the experimental environment is more complicated, there is hardly any process with negligible backgrounds, and calculations are generically less precise.

This means that at the LHC we need to consider two kinds of processes. The first involves all known particles, like electrons or W and Z bosons, or the top quark, or even the Higgs boson. These processes we call backgrounds, and they are described by QCD. The Higgs boson is in the middle of a transition to a background, only a few years ago is was the most famous example for a signal. By definition, signals are very rare compared to backgrounds. As an example, Fig. 7.1 shows that at the LHC the production cross section for a pair of bottom quarks is larger than 10^5 nb or 10^{11} fb, the typical production rate for W or Z bosons ranges around 200 nb or $2 \cdot 10^8$ fb, the rate for a pair of 500 GeV supersymmetric gluinos would have been $4 \cdot 10^4$ fb.

One LHC aspect we have to mention in the context of dark matter searches is the trigger. At the LHC we can only save and study a small number of all events. This means that we have to decide very fast if an event has the potential of being interesting in the light of the physics questions we are asking at the LHC; only these events we keep. For now we can safely assume that above an energy threshold we will keep all events with leptons or photons, plus, if at all possible, events with missing energy, like neutrinos in the Standard Model and dark matter particles in new physics models and jets with high energy coming from resonance decays.

When we search for dark matter particles at hadron colliders like the LHC, these analyses cannot rely on our knowledge of the initial state kinematics. What we know is that in the transverse plane the incoming partons add to zero three-momentum. In contrast, we are missing the necessary kinematic information in the beam direction. This means that dark matter searches always rely on production with another particle, leading to un-balanced three-momenta in the plane transverse to the beam direction. This defines an observable missing transverse momentum

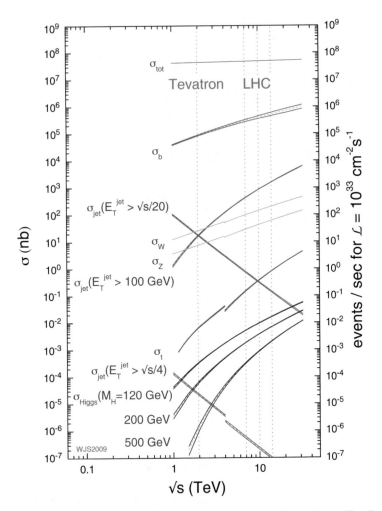

Fig. 7.1 Production rates for signal and background processes at hadron colliders. The disconti-nuity is due to the Tevatron being a proton–antiproton collider while the LHC is a proton–proton collider. The two colliders correspond to the *x*-axis values of 2 TeV and something between 7 and 14 TeV. Figure from Ref. [1]

three-vector with two relevant dimensions. The missing transverse energy is the absolute value of this two-dimensional three-vector. The big problem with missing transverse momentum is that it relies on reconstructing the entire recoil. This causes several experimental problems:

1. there will always be particles in events which are not observed in the calorime-ters. For example, a particle can hit a support structure of the detector, generating fake missing energy;

2. in particular hadronic jets might not be fully reconstructed, leading to fake missing energy in the direction of this jet. This is the reason why we usually require the missing momentum vector to not be aligned with any hard object in an event;
3. slight mis-measurements of the momenta of each of the particles in an event add, approximately, in quadrature to a mis-measurement of the missing energy vector;
4. QCD activity from the underlying event or from pile-up which gets subtracted before we analyze anything. This subtraction adds to the error on the measured missing momentum;
5. non-functional parts of the detector automatically lead to a systematic bias in the missing momentum distribution.

Altogether these effects imply that missing transverse energy below $30 \ldots 50\,\text{GeV}$ at the LHC could as well be zero. Only cuts on $E_{T,miss} \gtrsim 100\,\text{GeV}$ can guarantee a significant background rejection.

Next, we want to compute the production rates for dark matter particles at the LHC. To do that we need to follow the same path as for direct detection in Chap. 6, namely link the calculable partonic cross section to the observable hadronic cross section. We cannot compute the energy distributions of the incoming partons inside the colliding protons from first principles, but we can start with the assumption that all partons move collinearly with the surrounding proton. In that case the parton kinematics is described by a one-dimensional probability distribution for finding a parton just depending on the respective fraction of the proton's momentum, the parton density function (pdf) $f_i(x)$ with $x = 0 \ldots 1$ and $i = u, d, c, s, g$. This parton density itself is not an observable; it is a distribution in the mathematical sense, which means it is only defined when we integrate it together with a partonic cross section. Different parton densities have very different behavior—for the valence quarks (uud) they peak somewhere around $x \lesssim 1/3$, while the gluon pdf is negligible at $x \sim 1$ and grows very rapidly towards small x, $f_g(x) \propto x^{-2}$. Towards $x < 10^{-3}$ it becomes even steeper.

In addition, we can make some arguments based on symmetries and properties of the hadrons. For example the parton distributions inside an anti-proton are linked to those inside a proton through the CP symmetry, which is an exact symmetry of QCD,

$$f_q^{\bar{p}}(x) = f_{\bar{q}}(x), \qquad f_{\bar{q}}^{\bar{p}}(x) = f_q(x), \qquad f_g^{\bar{p}}(x) = f_g(x), \qquad (7.9)$$

for all x. The proton consists of uud quarks, plus quantum fluctuations either involving gluons or quark–antiquark pairs. The expectation values for up- and down-quarks have to fulfill

$$\int_0^1 dx \; (f_u(x) - f_{\bar{u}}(x)) = 2 \qquad \int_0^1 dx \; (f_d(x) - f_{\bar{d}}(x)) = 1 \, .$$

$$(7.10)$$

Finally, the proton momentum has to be the sum of all parton momenta, defining the
QCD sum rule

$$\langle \sum x_i \rangle = \int_0^1 dx \; x \left(\sum_q f_q(x) + \sum_{\bar{q}} f_{\bar{q}}(x) + f_g(x) \right) = 1 \; . \tag{7.11}$$

We can compute this sum accounting for quarks and antiquarks. The sum comes out
to 1/2, which means that half of the proton momentum is carried by gluons.

Using the parton densities we can compute the hadronic cross section,

$$\sigma_{\text{tot}} = \int_0^1 dx_1 \int_0^1 dx_2 \sum_{ij} f_i(x_1) \, f_j(x_2) \, \hat{\sigma}_{ij}(x_1 x_2 S) \; , \tag{7.12}$$

where i, j are the incoming partons. The partonic energy of the scattering process
is $s = x_1 x_2 S$ with the LHC proton energy of currently around $\sqrt{S} = 13\,\text{TeV}$. The
partonic cross section includes energy–momentum conservation.

On the parton level, the analogy to photon radiation in $e^+ e^-$ production will be
dark matter production together with a quark or a gluon. Two Feynman diagrams
for this mono-jet signature with an unspecified mediator are

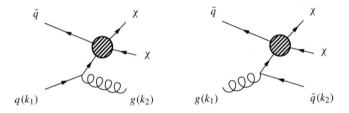

In addition to this experimental argument there is a theoretical, QCD argument
which suggests to look for initial state radiation of a quark or a gluon. Both of
the above diagrams include an intermediate quark or gluon propagator with the
denominator

$$\frac{1}{(k_1 - k_2)^2} = \frac{1}{k_1^2 - 2k_1^0 k_2^0 + 2(\vec{k}_1 \vec{k}_2) + k_2^2}$$

$$= \frac{1}{2} \frac{1}{|\vec{k}_1||\vec{k}_2| \cos\theta_{12} - k_1^0 k_2^0} = \frac{1}{2 k_1^0 k_2^0} \frac{1}{\cos\theta_{12} - 1} \; . \tag{7.13}$$

This propagator diverges when the radiated parton is soft ($k_2^0 \to 0$) or collinear
with the incoming parton ($\theta_{12} \to 0$). Phenomenologically, the soft divergence is
less dangerous, because the LHC experiments can only detect any kind of particle
above a certain momentum or transverse momentum threshold. The actual pole in
the collinear divergence gets absorbed into a re-definition of the parton densities

$f_{q,g}(x)$, as they appear for example in the hadronic cross section of Eq. (7.12). This so-called mass factorization is technically similar to a renormalization procedure for example of the strong coupling, except that renormalization absorbs ultraviolet divergences and works on the fundamental Lagrangian level [2]. One effect of this re-definition of the parton densities is that relative to the original definition the quark and gluon densities mix, which means that the two Feynman diagrams shown above cannot actually be separated on a consistent quantum level.

Experimentally, the scattering or polar angle θ_{12} is not the variable we actually measure. The reason is that it is not boost invariant and that we do not know the partonic rest frame in the beam direction. Instead, we can use two standard kinematic variables,

$$
t = -s \left(1 - \frac{m_{\chi\chi}^2}{s}\right) \frac{1 - \cos\theta_{12}}{2} \qquad \text{(Mandelstam variable)}
$$

$$
p_T^2 = s \left(1 - \frac{m_{\chi\chi}^2}{s}\right)^2 \frac{1 - \cos\theta_{12}}{2} \frac{1 + \cos\theta_{12}}{2} \qquad \text{(transverse momentum)} .
$$

$$(7.14)$$

Comparing the two forms we see that the transverse momentum is symmetric under the switch $\cos\theta_{12} \leftrightarrow -\cos\theta_{12}$, which in terms of the Mandelstam variables corresponds to $t \leftrightarrow u$. From Eq. (7.14) we see that the collinear divergence appears as a divergence of the partonic transverse momentum distribution,

$$
\frac{d\sigma_{\chi\chi j}}{dp_{T,j}} \propto |\mathcal{M}_{\chi\chi j}|^2 \propto \frac{1}{t} \propto \frac{1}{p_{T,j}^2} .
$$

$$(7.15)$$

An obvious question is whether this divergence is integrable, i.e. if it leads to a finite cross section $\sigma_{\chi\chi j}$. We can approximate the phase space integration in the collinear regime using an appropriate constant C to write

$$
\sigma_{\chi\chi j} \approx \int_{p_{T,j}^{\min}}^{p_{T,j}^{\max}} dp_{T,j}^2 \frac{C}{p_{T,j}^2} = 2 \int_{p_{T,j}^{\min}}^{p_{T,j}^{\max}} dp_{T,j} \frac{C}{p_{T,j}} = 2C \log \frac{p_{T,j}^{\max}}{p_{T,j}^{\min}} .
$$

$$(7.16)$$

For an integration of the full phase space including a lower limit $p_{T,j}^{\min} = 0$, this logarithm is divergent. When we apply an experimental cut to generate for example a value of $p_{T,j}^{\min} = 10\,\text{GeV}$, the logarithm gets large, because $p_{T,j}^{\max} \gtrsim 2m_\chi$ is given by the typical energy scales of the scattering process. When we absorb the collinear divergence into re-defined parton densities and use the parton shower to enforce and simulate the correct behavior

$$
\frac{d\sigma_{\chi\chi j}}{dp_{T,j}} \xrightarrow{p_{T,j} \to 0} 0 ,
$$

$$(7.17)$$

the large collinear logarithm in Eq. (7.16) gets re-summed to all orders in perturbation theory. However, over a wide range of values the transverse momentum distribution inherits the collinearly divergent behavior. This means that most jets radiated from incoming partons appear at small transverse momenta, and even after including the parton shower regulator the collinear logarithm significantly enhances the probability to radiate such collinear jets. The same is (obviously) true for the initial state radiation of photons. The main difference is that for the photon process we can neglect the amplitude with an initial state photon due to the small photon parton density.

Once we know that at the LHC we can generally look for the production of dark matter particles with an initial state radiation object, we can study different mono-X channels. Some example Feynman diagrams for mono-jet, mono-photon, and mono-Z production are

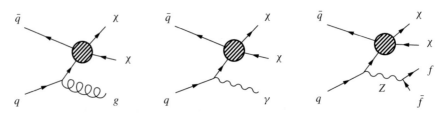

For the radiated Z-boson we need to specify a decay. While hadronic decays $Z \to q\bar{q}$ come with a large branching ratio, we need to ask what they add to the universal mono-jet signature. Leptonic decays like $Z \to \mu\mu$ can help in difficult experimental environments, but are suppressed by a branching ratio of 3.4% per lepton generation. Mono-W events can occur through initial state radiation when we use a $q\bar{q}'$ initial state to generate a hard $q\bar{q}$ scattering. Finally, mono-Higgs signatures obviously make no sense for initial state radiation. From the similarity of the above Feynman diagrams we can first assume that at least in the limit $m_Z \to 0$ the total rates for the different mono-X processes relative to the mono-jet rate scale like

$$\frac{\sigma_{\chi\chi\gamma}}{\sigma_{\chi\chi j}} \approx \frac{\alpha}{\alpha_s} \frac{Q_q^2}{C_F} \approx \frac{1}{40}$$

$$\frac{\sigma_{\chi\chi\mu\mu}}{\sigma_{\chi\chi j}} \approx \frac{\alpha}{\alpha_s} \frac{Q_q^2 s_w^2}{C_F} \ \mathrm{BR}(Z \to \mu\mu) \approx \frac{1}{4000} . \tag{7.18}$$

The actual suppression of the mono-Z channel is closer to 10^{-4}, once we include the Z-mass suppression through the available phase space. In addition, the similar Feynman diagrams also suggest that any kinematic x-distribution scales like

$$\frac{1}{\sigma_{\chi\chi j}} \frac{d\sigma_{\chi\chi g}}{dx} \approx \frac{1}{\sigma_{\chi\chi\gamma}} \frac{d\sigma_{\chi\chi\gamma}}{dx} \approx \frac{1}{\sigma_{\chi\chi ff}} \frac{d\sigma_{\chi\chi ff}}{dx} . \tag{7.19}$$

Here, the suppression of the mono-photon is stronger, because the rapidity coverage of the detector for jets extends to $|\eta| < 4.5$, while photons rely on an efficient electromagnetic calorimeter with $|\eta| < 2.5$. On the other hand, photons can be detected to significantly smaller transverse momenta than jets.

Note that the same scaling as in Eq. (7.18) applies to the leading mono-X backgrounds, namely

$$pp \rightarrow Z_{\nu\nu} X \qquad \text{with} \quad X = j, \gamma, Z\,, \qquad (7.20)$$

possibly with the exception of mono-Z production, where the hard process and the collinear radiation are now both described by Z-production. This means that the signal scaling of Eq. (7.18) also applies to backgrounds,

$$\frac{\sigma_{\nu\nu\gamma}}{\sigma_{\nu\nu j}} \approx \frac{\alpha}{\alpha_s} \frac{Q_q^2}{C_F} \approx \frac{1}{40}$$

$$\frac{\sigma_{\nu\nu\mu\mu}}{\sigma_{\nu\nu j}} \approx \frac{\alpha}{\alpha_s} \frac{Q_q^2 \, s_w^2}{C_F} \, \text{BR}(Z \rightarrow \mu\mu) \approx \frac{1}{4000}\,. \qquad (7.21)$$

If our discovery channel is statistics limited, the significances n_σ for the different channels are given in terms of the luminosity, efficiencies, and the cross sections

$$n_{\sigma,j} = \sqrt{\epsilon_j \mathcal{L}} \, \frac{\sigma_{\chi\chi j}}{\sqrt{\sigma_{\nu\nu j}}} \Rightarrow n_{\sigma,\gamma} = \sqrt{\epsilon_\gamma \mathcal{L}} \, \frac{\sigma_{\chi\chi\gamma}}{\sqrt{\sigma_{\nu\nu\gamma}}}$$

$$\approx \sqrt{\epsilon_j \mathcal{L}} \, \frac{1}{\sqrt{40}} \sqrt{\frac{\epsilon_\gamma}{\epsilon_j}} \, \frac{\sigma_{\chi\chi j}}{\sqrt{\sigma_{\nu\nu j}}} = \frac{1}{6.3} \sqrt{\frac{\epsilon_\gamma}{\epsilon_j}} \, n_{\sigma,j}\,. \qquad (7.22)$$

Unless the efficiency correction factors, including acceptance cuts and cuts rejecting other backgrounds, point towards a very significant advantage if the mono-photon channel, the mono-jet channel will be the most promising search strategy. Using the same argument, the factor in the expected mono-jet and mono-Z significances will be around $\sqrt{6000} = 77$.

This estimate might change if the uncertainties are dominated by systematics or a theory uncertainty. These errors scale proportional to the number of background events in the signal region, again with a signature-dependent proportionality factor ϵ describing how well we know the background distributions. This means for the significances

$$n_{\sigma,\gamma} = \epsilon_\gamma \frac{\sigma_{\chi\chi\gamma}}{\sigma_{\nu\nu\gamma}} = \frac{\epsilon_\gamma}{\epsilon_j} \, \epsilon_j \frac{\sigma_{\chi\chi j}}{\sigma_{\nu\nu j}} = \frac{\epsilon_\gamma}{\epsilon_j} \, n_{\sigma,j}\,. \qquad (7.23)$$

Typically, we understand photons better than jets, both experimentally and theoretically. On the other hand, systematic and theory uncertainties at the LHC are usually limited by the availability and the statistics in control regions, regions which we can safely assume to be described by the Standard Model.

We can simulate mono-X signatures for vector mediators, described in Sect. 5.4. In that case the three mono-X signatures are indeed induced by initial state radiation. The backgrounds are dominated by Z-decays to neutrinos. The corresponding LHC searches are based on the missing transverse momentum distribution and the transverse momentum $p_{T,X}$ of the mono-X object. There are (at least) two strategies to control for example the mono-jet background: first, we can measure it for example using $Z \to \mu^+\mu^-$ decays or hard photons produced in association with a hard jet. Second, if the dark matter signal is governed by a harder energy scale, like the mass of a heavy mediator, we can use the low-p_T region as a control region and only extrapolate the p_T distributions.

Figure 7.2 gives an impression of the transverse momentum spectra in the mono-jet, mono-photon, and mono-Z channels. Comparing the mono-jet and mono-photon rates we see that the shapes of the transverse momentum spectra of the jet or photon, recoiling against the dark matter states, are essentially the same in both cases, for the respective signals as well as for the backgrounds. The signal and background rates follow the hierarchy derived above. Indeed, the mono-photon hardly adds anything to the much larger mono-jet channel, except for cases where in spite of advanced experimental strategies the mono-jet channel is limited by systematics. The mono-Z channel with a leptonic Z-decay is kinematically almost identical to the other two channels, but with a strongly reduced rate. This means that for mono-X signatures induced by initial state radiation the leading mono-

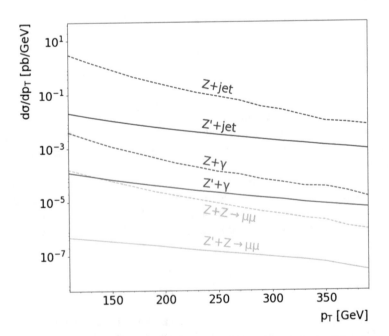

Fig. 7.2 Transverse momentum spectrum for signals and backgrounds in the different mono-X channels for a heavy vector mediator with $m_{Z'} = 1$ TeV. Figure from Ref. [3]

jet channel can be expected to be the most useful, while other mono-X analyses will only become interesting when the production mechanism is not initial state radiation.

Finally, one of the main challenges of mono-X signatures is that by definition the mediator has to couple to the Standard Model and to dark matter. This means for example in the case of the simple model of Eq. (5.27)

$$\mathrm{BR}(V \to q\bar{q}) + \mathrm{BR}(V \to \chi\chi) = 100\% . \tag{7.24}$$

The relative size of the branching ratios is given by the ratio of couplings g_χ^2/g_u^2. Instead of the mono-X signature we can constrain part of the model parameter space through resonance searches with the same topology as the mono-X search and without requiring a hard jet,

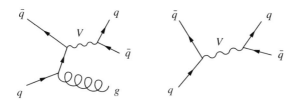

On the other hand, for the parameter space $g_u \ll g_\chi$ but constant $g_u g_\chi$ and mediator mass, the impact of resonance searches is reduced, whereas mono-X searches remain relevant.

7.3 Higgs Portal

In addition to the very general mono-jet searches for dark matter, we will again look at our two specific models. The Higgs portal model only introduces one more particle, a heavy scalar with $m_S \gg m_H$ and only coupling to the Higgs. This means that the Higgs has to act as an s-channel mediator not only for dark matter annihilation, but also for LHC production,

$$pp \to H^* \to SS + \mathrm{jets} . \tag{7.25}$$

The Higgs couples to gluons in the incoming protons through a top loop, which implies that its production rate is very small. The Standard Model predicts an on-shell Higgs rate of 50 pb for gluon fusion production at a 14 TeV LHC. Alternatively, we can look for weak-boson-fusion off-shell Higgs production, i.e. production in association with two forward jets. The corresponding Feynman diagram is

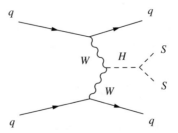

These so-called tagging jets will allow us to trigger the events. For an on-shell Higgs boson the weak boson fusion cross section at the LHC is roughly a factor 1/10 below gluon fusion, and its advantages are discussed in detail in Ref. [2].

In particular in this weak-boson-fusion channel ATLAS and CMS are conducting searches for invisibly decaying Higgs bosons. The main backgrounds are invisible Z-decays into a pair of neutrinos, and W-decays where we miss the lepton and are only left with one neutrino. For high luminosities around $3000\,\mathrm{fb}^{-1}$ and assuming an essentially unchanged Standard Model Higgs production rate, the LHC will be sensitive to invisible branching ratios around

$$\mathrm{BR}(H \to \text{invisible}) = (2\text{--}3)\% \ . \tag{7.26}$$

The key to this analysis is to understand not only the tagging jet kinematics, but also the central jet radiation between the two forward tagging jets.

Following the discussion in Sect. 4.1 the partial width for the SM Higgs boson decays into light dark matter is

$$\Gamma(H \to SS) = \frac{\lambda_3^2 v_H^2}{32\pi M_H} \sqrt{1 - \frac{4m_S^2}{m_H^2}} \quad\Leftrightarrow\quad \frac{\Gamma(H \to SS)}{m_H}$$

$$\approx \frac{\lambda_3^2}{8\pi} \left(1 - \frac{2m_S^2}{m_H^2} \right) < \frac{\lambda_3^2}{8\pi} \ . \tag{7.27}$$

This value has to be compared to the Standard Model prediction $\Gamma_H/m_H = 4 \cdot 10^{-5}$. For example, a 10% invisible branching ratio $\mathrm{BR}(H \to SS)$ into very light scalars $m_S \ll m_H/2$ corresponds to a portal coupling

$$\frac{\lambda_3^2}{8\pi} = 4 \cdot 10^{-6} \quad\Leftrightarrow\quad \lambda_3 = \sqrt{32\pi} \cdot 10^{-3} \approx 10^{-2} \ . \tag{7.28}$$

The light scalar reference point in agreement with the observed relic density Eq. (4.14) has $\lambda_3 = 0.3$ and roughly assuming $m_S \lesssim 50\,\mathrm{GeV}$. This is well above the approximate final reach for the invisible Higgs branching ratio at the high-luminosity LHC.

For larger dark matter masses above $m_S = 200\,\text{GeV}$ the LHC cross section for pair production in weak boson fusion is tiny, namely

$$\sigma(SSjj) \approx \frac{\lambda_3^2}{10}\,\text{fb} \overset{\lambda_3=0.1}{=} 10^{-3}\,\text{fb} \tag{7.29}$$

Without going into much detail this means that heavy scalar dark matter is unlikely to be discovered at the LHC any time soon, because the final state is heavy and the coupling to the Standard Model is strongly constrained through the observed relic density.

7.4 Supersymmetric Neutralinos

The main feature of supersymmetry is that it is not just a theory predicting a dark matter particle, it is a complete, renormalizable ultraviolet completion of the Standard Model valid to the Planck scale. From Sect. 4.3 we know that the MSSM and the NMSSM offer a wide variety of particles, including messengers linking the visible matter and dark matter sectors. Obviously, the usual mono-X signatures from Sect. 7.2 or the invisible Higgs decays from Sect. 7.3 will also appear in supersymmetric models. For example SM-like Higgs decays into a pair of light neutralinos can occur for a mixed gaugino-higgsino LSP with $M_1 \lesssim |\mu| \lesssim 100\,\text{GeV}$. An efficient annihilation towards the observed relic density goes through an s-channel Z-mediator coupling to the higgsino fraction. Here we can find

$$\text{BR}(h \to \tilde{\chi}_1^0 \tilde{\chi}_1^0) = (10 \ldots 50)\% \quad m_{\tilde{\chi}_1^0} = (35 \ldots 40)\,\text{GeV and } (50 \ldots 55)\,\text{GeV}\,, \tag{7.30}$$

mostly constrained by direct detection. On the other hand, supersymmetric models offer many more dark matter signatures and provide a UV completion to a number of different simplified models. They are often linked to generic features of heavy new particle production, which is what we will focus on below.

If our signature consists of a flexible number of visible and invisible particles we rely on global observables. The visible mass is based on the assumption that we are looking for the decay of two heavy new states, where the parton densities will ensure that these two particles are produced close to threshold. We can then approximate the partonic energy $\sqrt{\hat{s}} \sim m_1 + m_2$ by some kind of visible energy. Without taking into account missing energy and just adding leptons ℓ and jets j the visible mass looks like

$$m_{\text{visible}}^2 = \left[\sum_{\ell,j} E\right]^2 - \left[\sum_{\ell,j} \vec{p}\right]^2 . \tag{7.31}$$

Similarly, Tevatron and LHC experiments have for a long time used an effective transverse mass scale which is usually evaluated for jets only, but can trivially be extended to leptons,

$$H_T = \sum_{\ell,j} E_T = \sum_{\ell,j} p_T \,, \tag{7.32}$$

assuming massless final state particles. In an alternative definition of H_T we sum over a number of jets plus the missing energy and skip the hardest jet in this sum. Obviously, we can add the missing transverse momentum to this sum, giving us

$$m_{\text{eff}} = \sum_{\ell,j,\text{miss}} E_T = \sum_{\ell,j,\text{miss}} p_T \,. \tag{7.33}$$

This effective mass is known to trace the mass of the heavy new particles decaying for example to jets and missing energy. This interpretation relies on the non-relativistic nature of the production process and our confidence that all jets included are really decay jets.

In the Standard Model the neutrino produces such missing transverse energy, typically through the decays $W \to \ell^+ \nu$ and $Z \to \nu\bar{\nu}$. In $W+$ jets events we can learn how to reconstruct the W mass from one observed and one missing particle. We construct a transverse mass in analogy to an invariant mass, but neglecting the longitudinal momenta of the decay products

$$
\begin{aligned}
m_T^2 &= \left(E_{T,\text{miss}} + E_{T,\ell}\right)^2 - \left(\vec{p}_{T,\text{miss}} + \vec{p}_{T,\ell}\right)^2 \\
&= m_\ell^2 + m_{\text{miss}}^2 + 2\left(E_{T,\ell}E_{T,\text{miss}} - \vec{p}_{T,\ell} \cdot \vec{p}_{T,\text{miss}}\right) \,,
\end{aligned} \tag{7.34}
$$

in terms of a transverse energy $E_T^2 = \vec{p}_T^2 + m^2$. By definition, it is invariant under—or better independent of—longitudinal boosts. Moreover, as the projection of the invariant mass onto the transverse plane it is also invariant under transverse boosts. The transverse mass is always smaller than the actual mass and reaches this limit for a purely transverse momentum direction, which means that we can extract m_W from the upper endpoint in the $m_{T,W}$ distribution. To reject Standard Model backgrounds we can simply require $m_T > m_W$.

The first supersymmetric signature we discuss makes use of the fact that already the neutralino-chargino sector involves six particles, four neutral and two charged. Two LHC processes reflecting this structure are

$$
\begin{aligned}
pp &\to \tilde{\chi}_2^0 \tilde{\chi}_1^0 \to (\ell^+ \ell^- \tilde{\chi}_1^0) \, \tilde{\chi}_1^0 \\
pp &\to \tilde{\chi}_1^+ \tilde{\chi}_1^- \to (\ell^+ \nu_\ell \tilde{\chi}_1^0) \, (\ell^- \bar{\nu}_\ell \tilde{\chi}_1^0) \,.
\end{aligned} \tag{7.35}
$$

The leptons in the decay of the heavier neutralinos and charginos can be replaced by other fermions. Kinematically, the main question is if the fermions arise from on-shell gauge bosons or from intermediate supersymmetric scalar partners of the leptons. The corresponding Feynman diagrams for the first of the two above processes are

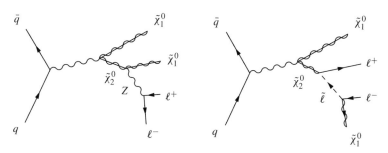

The question which decay topologies of the heavier neutralino dominate, depends on the point in parameter space. The first of the two diagrams predicts dark matter production in association with a Z-boson. This is the same signature as found to be irrelevant for initial state radiation in Sect. 7.2, namely mono-Z production.

The second topology brings us the question how many masses we can extract from two observed external momenta. Endpoint methods rely on lower (threshold) and upper (edge) kinematic endpoints of observed invariant mass distributions. The art is to identify distributions where the endpoint is probed by realistic phase space configurations. The most prominent example is $m_{\ell\ell}$ in the heavy neutralino decay in Eq. (7.35), proceeding through an on-shell slepton. In the rest frame of the intermediate slepton the $2 \rightarrow 2$ process corresponding to the decay of the heavy neutralino,

$$\tilde{\chi}_2^0 \ell^- \rightarrow \tilde{\ell} \rightarrow \tilde{\chi}_1^0 \ell^- \tag{7.36}$$

resembles the Drell–Yan process. Because of the scalar in the s-channel, angular correlations do not influence the $m_{\ell\ell}$ distribution, so it will have a triangular shape. Its upper limit or edge can be computed in the slepton rest frame. The incoming and outgoing three-momenta have the absolute values

$$|\vec{p}| = \frac{|m_{\tilde{\chi}_{1,2}^0}^2 - m_{\tilde{\ell}}^2|}{2m_{\tilde{\ell}}} , \tag{7.37}$$

assuming $m_\ell = 0$. The invariant mass of the two leptons reaches its maximum if the two leptons are back-to-back and the scattering angle is $\cos\theta = -1$

$$m_{\ell\ell}^2 = (p_{\ell^+} + p_{\ell^-})^2$$
$$= 2\left(E_{\ell^+} E_{\ell^-} - |\vec{p}_{\ell^+}||\vec{p}_{\ell^-}|\cos\theta\right)$$

$$< 2 \, (E_{\ell^+} E_{\ell^-} + |\vec{p}_{\ell^+}||\vec{p}_{\ell^-}|)$$

$$= 4 \, \frac{m_{\tilde{\chi}_2^0}^2 - m_{\tilde{\ell}}^2}{2 m_{\tilde{\ell}}} \, \frac{m_{\tilde{\ell}}^2 - m_{\tilde{\chi}_1^0}^2}{2 m_{\tilde{\ell}}} \quad \text{using} \quad E_{\ell^\pm}^2 = \vec{p}_{\ell^\pm}^2 \; . \tag{7.38}$$

The kinematic endpoint is then given by

$$0 < m_{\ell\ell}^2 < \frac{(m_{\tilde{\chi}_2^0}^2 - m_{\tilde{\ell}}^2)(m_{\tilde{\ell}}^2 - m_{\tilde{\chi}_1^0}^2)}{m_{\tilde{\ell}}^2} \; . \tag{7.39}$$

A generic feature or all methods relying on decay kinematics is that it is easier to constrain the differences of squared masses than the absolute mass scale. This is because of the form of the endpoint formulas, which involve the difference of mass squares $m_1^2 - m_2^2 = (m_1 + m_2)(m_1 - m_2)$. This combination is much more sensitive to $(m_1 - m_2)$ than it is to $(m_1 + m_2)$. The common lore that kinematics only constrain mass differences is not true for two body decays, but mass differences are indeed easier.

The second set of supersymmetric dark matter signatures involves the same extended dark matter sector with its neutralino and chargino spectra or a slepton. Because the slepton and the chargino are electrically charged, they can be produced through a photon mediator,

$$pp \to \tilde{\ell}\tilde{\ell}^* \quad \to (\ell^- \tilde{\chi}_1^0) \, (\ell^+ \tilde{\chi}_1^0)$$

$$pp \to \tilde{\chi}_1^+ \tilde{\chi}_1^- \to (\pi^+ \tilde{\chi}_1^0) \, (\pi^- \tilde{\chi}_1^0) \; . \tag{7.40}$$

For the slepton case one of the corresponding Feynman diagrams is

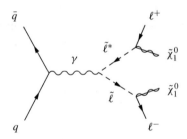

Again, the question arises how many masses we can extract from the measured external momenta. For this topology the variable m_{T2} generalizes the transverse mass known from W decays to the case of two massive invisible particles, one from each leg of the event. First, we divide the observed missing energy in the event into two scalar fractions $p_{T,\text{miss}} = q_1 + q_2$. Then, we construct the transverse mass for each side of the event, assuming that we know the invisible particle's mass or scanning over hypothetical values $\hat{m}_{\text{miss}}$.

Inspired by the transverse mass in Eq. (7.34) we are interested in a mass variable with a well-defined upper endpoint. For this purpose we construct some kind of minimum of $m_{T,j}$ as a function of the fractions q_j. We know that maximizing the transverse mass on one side of the event will minimize it on the other side, so we define

$$m_{T2}(\hat{m}_{\text{miss}}) = \min_{p_{T,\text{miss}}=q_1+q_2} \left[\max_j \; m_{T,j}(q_j; \hat{m}_{\text{miss}}) \right] . \qquad (7.41)$$

We can show that by construction the transverse mass fulfills

$$m_{\text{light}} + \hat{m}_{\text{miss}} < m_{T2}(\hat{m}_{\text{miss}})$$

$$m_{\text{light}} + m_{\text{miss}} < m_{T2}(m_{\text{miss}}) < m_{\text{heavy}} . \qquad (7.42)$$

For the correct value of m_{miss} the m_{T2} distribution has a sharp edge at the mass of the decaying particle. In favorable cases m_{T2} allows the measurement of both, the decaying particle and the invisible particle masses. These two aspects for the correct value $\hat{m}_{\text{miss}} = m_{\text{miss}}$ we can see in Fig. 7.3: the lower threshold is indeed given by $m_{T2} - m_{\tilde{\chi}_1^0} = m_\pi$, while the upper edge of $m_{T2} - m_{\tilde{\chi}_1^0}$ coincides with the dashed line for $m_{\tilde{\chi}_1^+} - m_{\tilde{\chi}_1^0}$.

An interesting aspect of m_{T2} is that it is boost invariant if and only if $\hat{m}_{\text{miss}} = m_{\text{miss}}$. For a wrong assignment of m_{miss} the value of m_{T2} has nothing to do with the actual kinematics and hence with any kind of invariant (and house numbers are not boost invariant). We can exploit this aspect by scanning over m_{miss} and looking for so-called kinks, defined as points where different events kinematics all return the same value for m_{T2}.

Finally, we can account for the fact that supersymmetry predicts new strongly interacting particles. These are the scalar partners of the quarks and the fermionic partner of the gluon. For dark matter physics the squarks are more interesting,

Fig. 7.3 Simulations for the decay $\tilde{\chi}_1^+ \to \tilde{\chi}_1^0 \pi$ or $\tilde{\chi}_1^+ \to \tilde{\chi}_1^0 e^+ \nu$. The blue m_{T2} line applies to the two-body decay. Figure from Ref. [4]

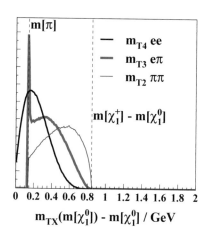

because their quark-squark-neutralino coupling makes them dark matter mediators to the strongly interacting visible matter sector. The same coupling also allows the squarks to decay into dark matter and a jet, leading to the dark matter signature

$$pp \to \tilde{q}\tilde{q}^* \to (q\tilde{\chi}_1^0)\,(\bar{q}\tilde{\chi}_1^0) \tag{7.43}$$

Example Feynman diagrams showing the role of squarks as t-channel colored mediators and as heavy particles decaying to dark matter are

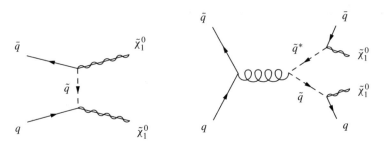

Note that these two squark-induced signatures cannot be separated, because they rely on the same two couplings, the quark-squark-neutralino coupling and the QCD-induced squark coupling to a gluon. Kinematically, they add nothing new to the above arguments: the first diagram will contribute to the mono-jet signature, with the additional possibility to radiate a gluon off the t-channel mediator, and to pair-production of neutralinos and charginos; the second diagram asks for a classic m_{T2} analysis. Moreover, the production process of Eq. (7.43) is QCD-mediated and the 100% branching fraction gives us no information about the mediator interaction to dark matter. In other words, for this pair-production process there exists no link between LHC observables and dark matter properties.

The non-negligible effect of the t-channel squark mediator adding to the s-channel Z-mediator for processes of the kind

$$pp \to \tilde{\chi}_i^0 \tilde{\chi}_j^0 \tag{7.44}$$

has to do with the couplings. From Eq. (4.43) we know that for neutralinos the higgsino content couples to the Z-mediator while the gaugino content couples to light-flavor squarks. In addition, the s-channel and t-channel diagrams typically interfere destructively, so we can tune the squark mass to significantly reduce the neutralino pair production cross section. The largest cross section for direct neutralino-chargino production is usually

$$pp \to \tilde{\chi}_1^+ \tilde{\chi}_2^0 \to (\ell^+ \nu \tilde{\chi}_1^0)\,(\ell^+ \ell^- \tilde{\chi}_1^0) \quad \text{with} \quad \sigma(\tilde{\chi}_1^\pm \tilde{\chi}_2^0) \lesssim 1\ \text{pb}\,, \tag{7.45}$$

for $m_\chi > 200\,\text{GeV}$. This decay leads to a tri-lepton signature with off-shell gauge bosons in the decay. The backgrounds are pair production of weak bosons and hence small. Just as a comparison, squark pair production, $pp \to \tilde{q}\tilde{q}^*$, can reach cross sections in the pico-barn range even for squark mass above 1 TeV.

Before the LHC started running, studies of decay chains with dark matter states at their end were in fashion. Here, squark decays had large impact through the stereotypical cascade decay

$$\tilde{q} \to \tilde{\chi}_2^0 \, q \to \tilde{\ell}^{\pm} \ell \, q \to \tilde{\chi}_1^0 \ell^+ \ell^- \, q \; . \tag{7.46}$$

First, we need to remove top pair production as the main background for such signatures. The key observation is that in cascade decays the leptons are flavor-locked, which means the combination $e^+ e^- + \mu^+ \mu^- - e^- \mu^+ - e^+ \mu^-$ is roughly twice $\mu^+ \mu^-$ for the signal, while it cancels for top pairs. In addition, such cascade decays are an opportunity to search for kinematic endpoints in many distributions, like $m_{\ell\ell}$, $m_{q\ell}$, or three-body combinations. Unfortunately, the general interest in the kinematics of supersymmetric cascade decays is for now postponed.

One thing we know for example from the di-lepton edge is that invariant masses can just be an invariant way of writing angular correlations between outgoing particles. Those depend on the spin and quantum numbers of all particles involved. While measuring for example the spin of new particles is hard in the absence of fully reconstructed events, we can try to probe it in the kinematics of cascade decays. The squark decay chain was the first case where such a strategy was worked out [5]:

1. Instead of measuring individual spins in a cascade decay we assume that cascade decays radiate particles with known spins. For radiated quarks and leptons the spins inside the decay chain alternate between fermions and bosons. Therefore, we contrast supersymmetry with another hypothesis, where the spins in the decay chain follow the Standard Model assignments. An example for such a model are Universal Extra Dimensions, where each Standard Model particle acquires a Kaluza–Klein partner from the propagation in the bulk of the additional dimensions;

2. The kinematical endpoints are completely determined by the masses and cannot be used to distinguish between the spin assignments. In contrast, the distributions between endpoints reflect angular correlations. For example, the $m_{j\ell}$ distribution in principle allows us to analyze spin correlations in squark decays in a Lorentz-invariant way. The only problem is the link between $\ell^{\pm}$ and their ordering in the decay chain;

3. As a proton–proton collider the LHC produces considerably more squarks than anti-squarks in the squark–gluino production process. A decaying squark radiates a quark while an antisquark radiates an antiquark, which means that we can define a non-zero production-side asymmetry between $m_{j\ell^+}$ and $m_{j\ell^-}$. Such an asymmetry we show in Fig. 7.4, for the SUSY and for the UED hypotheses. Provided the masses in the decay chain are not too degenerate we can indeed distinguish the two hypotheses.

Fig. 7.4 Asymmetry in $m_{j\ell}/m_{j\ell}^{max}$ for supersymmetry (dashed) and universal extra dimensions (solid). The spectrum is assumed to be hierarchical, which is typical for supersymmetric theories. Figure from Ref. [5]

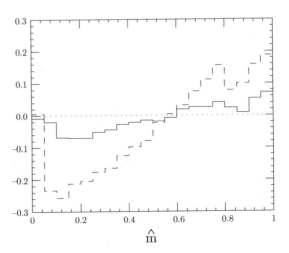

7.5   Effective Field Theory

In Sect. 4.4 we introduced an effective field theory of dark matter to describe dark matter annihilation in the early universe. If the annihilation process is the usual $2 \rightarrow 2$ WIMP scattering process it is formulated in terms of a dark matter mass m_χ and a mediator mass m_{med}, where the latter does not correspond to a propagating degree of freedom. It can hence be identified with a general suppression scale Λ in an effective Lagrangian, like the one illustrated in Eq. (4.59). All experimental environments discussed in the previous sections, including the relic density, indirect detection, and direct detection, rely on non-relativistic dark matter scattering. This means they can be described by a dark matter EFT if the mediator is much heavier than the dark matter agent,

$$\boxed{m_\chi \ll m_{\mathrm{med}}} . \tag{7.47}$$

In contrast, LHC physics is entirely relativistic and neither the incoming partons nor the outgoing dark matter particles in the schematic diagram shown in Sect. 1.3 are at low velocities. This means we have to add the partonic energy of the scattering process to the relevant energy scales,

$$\{ m_\chi, m_{\mathrm{med}}, \sqrt{s} \} . \tag{7.48}$$

In the case of mono-jet production, described in Sect. 7.2, the key observables here are the $\not{E}_T$ and $p_{T,j}$ distributions. For simple hard processes the two transverse momentum distributions are rapidly dropping and strongly correlated. This defines the relevant energy scales as

$$\{ m_\chi, m_{\mathrm{med}}, \not{E}_T^{\mathrm{min}} \} . \tag{7.49}$$

The experimentally relevant $\not{E}_T$ or $p_{T,j}$ regime is given by a combination of the signal mass scale, the kinematics of the dominant $Z_{\nu\nu}$+jets background, and triggering. Our effective theory then has to reproduce two key observables,

$$\sigma_{\text{tot}}(m_\chi, m_{\text{med}})\Big|_{\text{acceptance}} \quad \text{and} \quad \frac{d\,\sigma(m_\chi, m_{\text{med}})}{d\,\not{E}_T} \sim \frac{d\,\sigma(m_\chi, m_{\text{med}})}{d\,p_{T,j}}. \tag{7.50}$$

For the total rate, different phase space regions which individually agree poorly between the effective theory and some underlying model, might combine to a decent rate. For the main distributions this is no longer possible.

Finally, the hadronic LHC energy of 13 TeV, combined with reasonable parton momentum fractions defines an absolute upper limit, above which for example a particle in the s-channel cannot be produced as a propagating state,

$$\boxed{\{m_\chi, m_{\text{med}}, \not{E}_T^{\text{min}}, \sqrt{s}_{\text{max}}\}}. \tag{7.51}$$

This fourth scale is not the hadronic collision energy 13 TeV. From the typical LHC reach for heavy resonances in the s-channel we expect it to be in the range $\sqrt{s}_{\text{max}} = 5 \ldots 8$ TeV, depending on the details of the mediator.

From what we know from these lecture notes, establishing a completely general dark matter EFT approach at the LHC is not going to work. The Higgs portal results of Sect. 7.3 indicate that the only way to systematically search for its dark matter scalar is through invisible Higgs decays. By definition, those will be entirely dominated by on-shell Higgs production, not described by an effective field theory with a non-propagating mediator. Similarly, in the MSSM a sizeable fraction of the mediators are either light SM particles or s-channel particles within the reach of the LHC. Moreover, we need to add propagating degrees of co-annihilation partners, more or less close to the dark matter sector.

On the other hand, the fact that some of our favorite dark matter models are not described well by an effective Lagrangian does not mean that we cannot use such an effective Lagrangian for other classes of dark matter models. One appropriate way to test the EFT approach at the LHC is to rely on specific simplified models, as introduced in Sect. 5.4. Three simplified models come to mind for a fermion dark matter agent [6]:

1. tree-level s-channel vector mediator, as discussed in Sect. 5.4;
2. tree-level t-channel scalar mediator, realized as light-flavor scalar quarks in the MSSM, Sect. 7.4;
3. loop-mediated s-channel scalar mediator, realized as heavy Higgses in the MSSM, Sect. 7.4.

For the tree-level vector the situation at the LHC already becomes obvious in Sect. 5.4. The EFT approach is only applicable when also at the LHC the vector mediator is produced away from its mass shell, requiring roughly $m_V > 5$ TeV. The problem in this parameter range is that the dark matter annihilation cross section will be typically too small to provide the observed relic density. This makes the parameter region where this mediator can be described by global EFT analyses very small.

We start our more quantitative discussion with a tree-level t-channel scalar $\tilde{u}$. Unlike for the vector mediator, the t-channel mediator model only makes sense in the half plane with $m_\chi < m_{\tilde{u}}$; otherwise the dark matter agent would decay. At the LHC we have to consider different production processes. Beyond the unobservable process $u\bar{u} \to \chi\chi$ the two relevant topologies leading to mono-jet production are

$$u\bar{u} \to \chi\bar{\chi}g \qquad \text{and} \qquad ug \to \chi\bar{\chi}u \;. \tag{7.52}$$

They are of the same order in perturbation theory and experimentally indistinguishable. The second process can be dominated by on-shell mediator production, $ug \to \chi\tilde{u} \to \chi\,(\bar{\chi}u)$. We can cross its amplitude to describe the co-annihilation process $\chi\tilde{u} \to ug$. The difference between the (co-) annihilation and LHC interpretations of the same amplitude is that it only contributes to the relic density for $m_{\tilde{u}} < m_\chi + 10\%$, while it dominates mono-jet production for a wide range of mediator masses.

Following Eq. (7.43) we can also pair-produce the necessarily strongly interacting mediators with a subsequent decay to two jets plus missing energy,

$$q\bar{q}/gg \xrightarrow{\text{QCD}} \tilde{u}\tilde{u}^* \xrightarrow{\text{dark matter}} (\bar{\chi}u)\,(\chi\bar{u}) \;. \tag{7.53}$$

The partonic initial state of this process can be quarks or gluons. For a wide range of dark matter and mediator masses this process completely dominates the $\chi\chi$+jets process.

When the t-channel mediator becomes heavy, for example mono-jet production with the partonic processes given in Eq. (7.52) can be described by an effective four-fermion operator,

$$\mathscr{L} \supset \frac{c}{\Lambda^2}\,(\bar{u}_R\chi)\,(\bar{\chi}u_R) \;. \tag{7.54}$$

The natural matching scale will be around $\Lambda = m_{\tilde{u}}$. Note that this operator mediates the t-channel as well as the single-resonant mediator production topologies and the pair-production process induced by quarks. In contrast, pair production from two gluons requires a higher-dimensional operator involving the gluon field strength, like for example

$$\mathscr{L} \supset \frac{c}{\Lambda^3}(\bar{\chi}\chi)\,G_{\mu\nu}G^{\mu\nu} \;. \tag{7.55}$$

This leads to a much faster decoupling pattern of the pair production process for a heavy mediator.

Because the t-channel mediator carries color charge, LHC constraints typically force us into the regime $m_{\tilde{u}} \gtrsim 1\,\text{TeV}$, where an EFT approach can be viable. In addition, we again need to generate a large dark matter annihilation rate, which based on the usual scaling can be achieved by requiring $m_{\tilde{u}} \gtrsim m_\chi$. For heavy mediators, pair production decouples rapidly and leads to a parameter region where single-resonant production plays an important role. It is described by the same effective Lagrangian as the generic t-channel process, and decouples more rapidly than the t-channel diagram for $m_{\tilde{u}} \gtrsim 5\,\text{TeV}$. These actual mass values unfortunately imply that the remaining parameter regions suitable for an EFT description typically predict very small LHC rates.

The third simplified model we discuss is a scalar s-channel mediator. To generate a sizeable LHC rate we do not rely on its Yukawa couplings to light quarks, but on a loop-induced coupling to gluons, in complete analogy to SM-like light Higgs production at the LHC. The situation is slightly different for most of the supersymmetric parameter space for heavy Higgses, which have reduced top Yukawa couplings and are therefore much harder to produce at the LHC. Two relevant Feynman diagrams for mono-jet production are

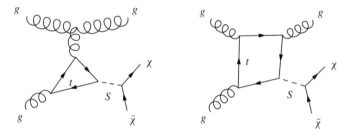

Coupling the scalar only to the top quark, we define the Lagrangian for the simplified scalar mediator model as

$$\mathscr{L} \supset -\frac{y_t m_t}{v}\, S\,\bar{t}t + g_\chi S\,\bar{\chi}\chi \tag{7.56}$$

The factor m_t/v in the top Yukawa coupling is conventional, to allow for an easier comparison to the Higgs case. The scalar coupling to the dark matter fermions can be linked to m_χ, but does not have to. We know that the SM Higgs is a very narrow resonance, while in this case the total width is bounded by the partial width from scalar decays to the top quark,

$$\frac{\Gamma_S}{m_S} > \frac{3G_F m_t^2 y_t^2}{4\sqrt{2}\pi}\left(1 - \frac{4m_t^2}{m_S^2}\right)^{3/2} \overset{m_S \gg m_t}{=} \frac{3G_F m_t^2 y_t^2}{4\sqrt{2}\pi} \approx 5\%\,, \tag{7.57}$$

assuming $y_t \approx 1$. Again, this is different from the case of supersymmetric, heavy Higgses, which can be broad.

To get a rough idea what kind of parameter space might be interesting, we can look at the relic density. The problem in this prediction is that for $m_\chi < m_t$ the annihilation channel $\chi\chi \to t\bar{t}$ is kinematically closed. Going through the same amplitude as the one for LHC production, very light dark matter will annihilate to two gluons through a top loop. If we allow for that coupling, the tree-level process $\chi\chi \to c\bar{c}$ will dominate for slightly heavier dark matter. If there also exists a Yukawa coupling of the mediator to bottom quarks, the annihilation channel $\chi\chi \to b\bar{b}$ will then take over for slightly heavier dark matter. An even heavier mediator will annihilate into off-shell top quarks, $\chi\chi \to (W^+b)(W^-\bar{b})$, and for $m_\chi > m_t$ the tree-level $2 \to 2$ annihilation process $\chi\chi \to t\bar{t}$ will provide very efficient annihilation. None of the aspects determining the correct annihilation channels are well-defined within the simplified model. Moreover, in the Lagrangian of Eq. (7.56) we can easily replace the scalar S with a pseudo-scalar, which will affect all non-relativistic processes.

For our global EFT picture this means that if a scalar s-channel mediator is predominantly coupled to up-quarks, the link between the LHC production rate and the predicted relic density essentially vanishes. The two observables are only related if the mediator is very light and decays through the one-loop diagram to a pair of gluons. This is exactly where the usual dark matter EFT will not be applicable.

If we only look at the LHC, the situation becomes much simpler. The dominant production process

$$gg \to S + \text{jets} \to \chi\chi + \text{jets} \tag{7.58}$$

defines the mono-jet signature through initial-state radiation and through gluon radiation off the top loop. The mono-jet rate will factorize into $\sigma_{S+j} \times \text{BR}_{\chi\chi}$. The production process is well known from Higgs physics, including the phase space region with a large jet and the logarithmic top mass dependence of the transverse momentum distribution,

$$\frac{d\sigma_{Sj}}{dp_{T,j}} = \frac{d\sigma_{Sj}}{dp_{T,S}} \propto \log^4 \frac{p_{T,j}^2}{m_t^2} . \tag{7.59}$$

Based on the Lagrangian given in Eq. (7.56) and the transverse momentum dependence given in Eq. (7.59), the mono-jet signal at the LHC depends on the four energy scales,

$$\{ m_\chi, m_S, m_t, \rlap{/}{E}_T = p_{T,j} \} , \tag{7.60}$$

which have to be organized in an effective field theory. If we focus on total rates, we are still left with three mass scales with different possible hierarchies:

1. The dark matter agent obviously has to remain a propagating degree of freedom, so in analogy to the SM Higgs case we can first assume a non-propagating top quark

$$m_t > m_S > 2m_\chi . \tag{7.61}$$

This defines the effective Lagrangian

$$\mathcal{L}^{(1)} \supset \frac{c}{\Lambda} S\, G_{\mu\nu} G^{\mu\nu} - g_\chi\, S\, \bar{\chi}\chi \,. \tag{7.62}$$

It is similar to the effective Higgs–gluon coupling in direct detection, defined in Eq. (6.14). The Wilson coefficient can be determined at the matching scale $\Lambda = m_t$ and assume the simple form

$$\frac{c}{\Lambda} \overset{m_t \gg m_S}{=} \frac{\alpha_s}{12\pi} \frac{y_t}{v} \,, \tag{7.63}$$

In this effective theory the transverse momentum spectra will fail to reproduce large logarithms of the type $\log(p_T/m_t)$, limiting the agreement between the simplified model and its EFT approximation.
2. Alternatively, we can decouple the mediator,

$$m_S > m_t, 2m_\chi \,, \tag{7.64}$$

leading to the usual dimension-6 four-fermion operators coupling dark matter to the resolved top loop,

$$\mathcal{L}^{(2)} \supset \frac{c}{\Lambda^2} (\bar{t}t)\, (\bar{\chi}\chi) \,. \tag{7.65}$$

The Wilson coefficients we obtain from matching at $\Lambda = m_S$ are

$$\frac{c}{\Lambda^2} = \frac{y_t g_\chi}{m_S^2} \frac{m_t}{v} \,. \tag{7.66}$$

This effective theory will retain all top mass effects in the distributions.
3. Finally, we can decouple the top as well as the mediator,

$$m_S, m_t > 2m_\chi \,. \tag{7.67}$$

The effective Lagrangian reads

$$\mathcal{L}^{(3)} \supset \frac{c}{\Lambda^3} (\bar{\chi}\chi)\, G_{\mu\nu} G^{\mu\nu} \,. \tag{7.68}$$

This dimension-seven operators is further suppressed by two equal heavy mass scales. Matching at $\Lambda = m_S \approx m_t$ gives us

$$\frac{c}{\Lambda^3} \overset{m_S \approx m_t \gg m_\chi}{=} \frac{\alpha_s}{12\pi} \frac{y_t g_\chi}{m_S^2} \frac{1}{v} \,, \tag{7.69}$$

assuming only the top quark runs in the loop.

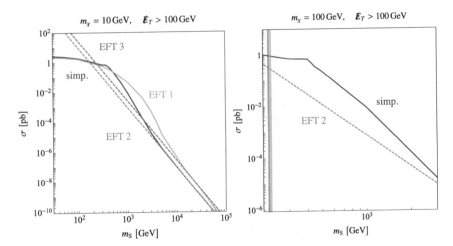

Fig. 7.5 Total mono-jet rate in the loop-mediated s-channel scalar model as function of the mediator mass for. We show all three different $m_\chi = 10$ GeV (left) and $m_\chi = 100$ GeV (right). For the shaded regions the annihilation cross section reproduces the observed relic density within $\Omega_\chi^{obs}/3$ and $\Omega_\chi^{obs} + 10\%$ for a mediator coupling only to up-type quarks (red) or to both types of quarks (green). Figure from Ref. [6]

We show the predictions for the total LHC rate based on these three effective theories and the simplified model in the left panel of Fig. 7.5. The decoupled top ansatz $\mathscr{L}^{(1)}$ of Eq. (7.62) indeed reproduces the correct total rate for $m_S < 2m_t$. Above that threshold it systematically overestimates the cross section. The effective Lagrangian $\mathscr{L}^{(2)}$ with a decoupled mediator, Eq. (7.65), reproduces the simplified model for $m_S \gtrsim 5$ TeV. Beyond this value the LHC energy is not sufficient to produce the mediator on-shell. Finally, the effective Lagrangian $\mathscr{L}^{(3)}$ with a simultaneously decoupled top quark and mediator, Eq. (7.68), does not reproduce the total production rate anywhere.

In the right panel of Fig. 7.5 we show the mono-jet rate for heavier dark matter and the parameter regions where the simplified model predicts a roughly correct relic density. In this range only the EFT with the decoupled mediator, defined in Eq. (7.65), makes sense. Because the model gives us this freedom, we also test what happens to the combination with the relic density when we couple the mediator to all quarks, rather than up-quarks only. Altogether, we find that in the region of heavy mediators the EFT is valid for LHC observables if

$$m_S > 5 \text{ TeV} . \tag{7.70}$$

This is similar to the range of EFT validity for the s-channel vector model.

References

1. Campbell, J.M., Huston, J.W., Stirling, W.J.: Hard interactions of quarks and gluons: a primer for LHC physics. Rep. Prog. Phys. **70**, 89 (2007). arXiv:hep-ph/0611148
2. Plehn, T.: Lectures on LHC Physics. Lect. Notes Phys. 886 (2015). arXiv:0910.4182 [hep-ph]. https://www.thphys.uni-heidelberg.de/~plehn/?visible=review
3. Bernreuther, E., Horak, J., Plehn, T., Butter, A.: Actual physics behind mono-X. SciPost. arXiv:1805.11637 [hep-ph]
4. Barr, A., Lester, C., Stephens, P.: m(T2): the truth behind the glamour. J. Phys. G **29**, 2343 (2003). arXiv:hep-ph/0304226
5. Smillie, J.M., Webber, B.R.: Distinguishing spins in supersymmetric and universal extra dimension models at the large hadron collider. J. High Energy Phys. **0510**, 069 (2005). arXiv:hep-ph/0507170
6. Bauer, M., Butter, A., Desai, N., Gonzalez-Fraile, J., Plehn, T.: Validity of dark matter effective theory. Phys. Rev. D **95**(7), 075036 (2017). arXiv:1611.09908 [hep-ph]

Chapter 8
Further Reading

First, we would like to emphasize that our list of references is limited to the, legally required, sources of figures and to slightly more advanced material providing more details about the topics discussed in these lecture notes.

Our discussion on the general relativity background and cosmology is a very brief summary. Dedicated textbooks include the classics by Kolb and Turner [1], Bergström and Goobar [2], Weinberg [3], as well as the more modern books by Dodelson [4] and Tegmark [5]. More details on the role of dark matter in the history of the universe is given in the book by Bertone and Hooper [6] and in the notes by Tanedo [7] and Mambrini [8]. Jim Cline's TASI lectures [9] serve as an up-to-date discussion on the role of dark matter in the history of the Universe. Further details on the cosmic microwave background and structure formation are also in the lecture notes on cosmological perturbation theory by Hannu Kurki-Suonio that are available online [10], as well as in the lecture notes on Cosmology by Rosa [11] and Baumann [12].

For models of particle dark matter, Ref. [13] provides a list of consistency tests. For further reading on WIMP dark matter we recommend the didactic review article Ref. [14]. Reference [15] addresses details on WIMP annihilation and the resulting constraints from the comic microwave background radiation. A more detailed treatment of the calculation of the relic density for a WIMP is given in Ref. [16]. Felix Kahlhöfer has written a nice review article on LHC searches for WIMPs [17]. For further reading on the effect of the Sommerfeld enhancement, we recommend Ref. [18].

Extensions of the WIMP paradigm can result in a modified freeze-out mechanism, as is the case of the co-annihilation scenario. These exceptions to the most straightforward dark matter freeze-out have originally been discussed by Griest and Seckel in Ref. [19]. A nice systematic discussion of recent research aiming can be found in Ref. [20].

© Springer Nature Switzerland AG 2019
M. Bauer, T. Plehn, *Yet Another Introduction to Dark Matter*,
Lecture Notes in Physics 959, https://doi.org/10.1007/978-3-030-16234-4_8

For models of non-WIMP dark matter, the review article Ref. [21] provides many details. A very good review of axions is given in Roberto Peccei's notes [22]. while axions as dark matter candidates are discussed in Ref. [23]. Mariangela Lisanti's TASI lectures [24] provide a pedagogical over these different dark matter candidates. Details on light dark matter, in particular hidden photons, can be found in Tongyan Lin's notes for her 2018 TASI lecture [25].

Details on calculations for the direct search for dark matter can be found in the review by Lewin and Smith [26]. Gondolo and Silk provide details for dark matter annihilation in the galactic center [27], as do the TASI lecture notes of Hooper [28]. For many more details on indirect detection of dark matter we refer to Tracy Slatyer's TASI lectures [29].

Note the one aspect these lecture notes are still missing is the chapter on the discovery of WIMPs. We plan to add an in-depth discussion of the WIMP discovery to an updated version of these notes.

References

1. Kolb, E.W., Turner, M.S.: The early universe. Front. Phys. **69**, 1 (1990)
2. Bergstrom, L., Goobar, A.: Cosmology and Particle Astrophysics. Chichester, Wiley (1999)
3. Weinberg, S.: Cosmology. Oxford University, Oxford (2008)
4. Dodelson, S.: Modern Cosmology. Academic, Amsterdam (2003)
5. Tegmark, M.: Measuring space-time: from big bang to black holes. In: The Early Universe and Observational Cosmology. Lect. Notes in Phys. 646, 169 (2004). arXiv:astro-ph/0207199
6. Bertone, G., Hooper, D.: A History of Dark Matter (2016). arXiv:1605.04909
7. Tanedo, F.: Defense Against the Dark Arts. http://www.physics.uci.edu/~tanedo/files/notes/DMNotes.pdf
8. Mambrini, Y.: Histories of Dark Matter in the Universe. http://www.ymambrini.com/My_World/Physics_files/Universe.pdf
9. Cline, J.M.: TASI Lectures on Early Universe Cosmology: Inflation, Baryogenesis and Dark Matter (2018). arXiv:1807.08749 [hep-ph]
10. Kurki-Suonio, H.: Cosmology I and II. http://www.helsinki.fi/~hkurkisu
11. Rosa, J.G.: Introduction to Cosmology. http://gravitation.web.ua.pt/cosmo
12. Baumann, D.: Cosmology. http://www.damtp.cam.ac.uk/people/d.baumann
13. Taoso, M., Bertone, G., Masiero, A.: Dark matter candidates: a ten-point test. J. Cosmol. Astropart. Phys. **0803**, 022 (2008). arXiv:0711.4996 [astro-ph]
14. Arcadi, G., Dutra, M., Ghosh, P., Lindner, M., Mambrini, Y., Pierre, M., Profumo, S., Queiroz, F.S.: The waning of the WIMP? A review of models, searches, and constraints. Eur. Phys. J. C **78**(3), 203 (2018). arXiv:1703.07364 [hep-ph]
15. Slatyer, T.R., Padmanabhan, N., Finkbeiner, D.P.: CMB constraints on WIMP annihilation: energy absorption during the recombination epoch. Phys. Rev. D **80**, 043526 (2009). arXiv:0906.1197 [astro-ph.CO]
16. Steigman, G., Dasgupta, B., Beacom, J.F.: Precise relic WIMP abundance and its impact on searches for dark matter annihilation. Phys. Rev. D **86**, 023506 (2012). arXiv:1204.3622 [hep-ph]
17. Kahlhoefer, F.: Review of LHC dark matter searches. Int. J. Mod. Phys. A **32**, 1730006 (2017). arXiv:1702.02430 [hep-ph]
18. Arkani-Hamed, N., Finkbeiner, D.P., Slatyer, T.R., Weiner, N.: A theory of dark matter. Phys. Rev. D **79**, 015014 (2009). arXiv:0810.0713 [hep-ph]

19. Griest, K., Seckel, D.: Three exceptions in the calculation of relic abundances. Phys. Rev. D **43**, 3191 (1991)
20. D'Agnolo, R.T., Pappadopulo, D., Ruderman, J.T.: Fourth exception in the calculation of relic abundances. Phys. Rev. Lett. **119**(6), 061102 (2017). arXiv:1705.08450 [hep-ph]
21. Baer, H., Choi, K.Y., Kim, J.E., Roszkowski, L.: Dark matter production in the early universe: beyond the thermal WIMP paradigm. Phys. Rep. **555**, 1 (2015). arXiv:1407.0017 [hep-ph]
22. Peccei, R.D.: The strong CP problem and axions. Lect. Notes Phys. 741, 3 (2008). arXiv:hep-ph/0607268
23. Arias, P., Cadamuro, D., Goodsell, M., Jaeckel, J., Redondo, J., Ringwald, A.: WISPy cold dark matter. J. Cosmol. Astropart. Phys. **1206**, 013 (2012). arXiv:1201.5902 [hep-ph]
24. Lisanti, M.: Lectures on Dark Matter Physics. arXiv:1603.03797 [hep-ph]
25. Lin, T.: Dark Matter Models and Direct Searches, Lecture at TASI 2018. Lecture Notes. https://www.youtube.com/watch?v=fQSWMsOfOcc
26. Lewin, J.D., Smith, P.F.: Review of mathematics, numerical factors, and corrections for dark matter experiments based on elastic nuclear recoil. Astropart. Phys. **6**, 87 (1996)
27. Gondolo, P., Silk, J.: Dark matter annihilation at the galactic center. Phys. Rev. Lett. **83**, 1719 (1999). arXiv:astro-ph/9906391
28. Hooper, D.: Particle Dark Matter (2009). arXiv:0901.4090 [hep-ph]
29. Slatyer, T.R.: Indirect Detection of Dark Matter (2017). arXiv:1710.05137 [hep-ph]

Index

© Springer Nature Switzerland AG 2019
M. Bauer, T. Plehn, *Yet Another Introduction to Dark Matter*,
Lecture Notes in Physics 959, https://doi.org/10.1007/978-3-030-16234-4